AF761149

INTEGRATED REVIEW WORKSHEETS

LAURA HILLERBRAND

Broward College

INTRODUCTORY STATISTICS: EXPLORING THE WORD THROUGH DATA

THIRD EDITION

Robert Gould

University of California, Los Angeles

Rebecca Wong

West Valley College

Colleen Ryan

Moorpark Community College

Pearson

Pearson

ISBN-13: 978-0-13-590078-9
ISBN-10: 0-13-590078-6

3 2026

Contents

Chapter 1 Introduction to Data

Objective 1: Identify variables in context.

A variable represents an unknown quantity. In statistics, this could be a data value. You may want to consider using a variable name that is helpful in identifying what it represents (e.g., w for weight, h for height).

Note that within this course, you will feel as though you are reading Greek at times because you will be! A lot of Greek letters are used as variables in this course. The good news is that if you ever travel to Greece, you will be able to read some of the signs there. To see some of the Greek alphabet that you may encounter in this course, please see Appendix A.

Example 1 Let n represent the number of respondents to a survey about exercise. What does $n = 50$ mean?

It means that 50 people responded to the survey about exercise.

My Turn!

Let h represent the height of a wave in feet at "Jaws" in Maui. What does $h = 30$ mean?

Example 2 A six-sided die is rolled.

a. Let d represent the side that faces up. What are the possible values for d?
   A six sided die has sides with 1, 2, 3, 4, 5, and 6 dots. So, d has the possible values of 1, 2, 3, 4, 5, or 6.

b. Let t represent the sides that face up that are multiples of 3. What are the possible values for t?
   The faces of a die that are multiples of 3 are 3 and 6. Therefore, t has the possible values of 3 or 6.

c. Now, the die is rolled 4 times. Let f represent the number of times that the die lands with a 5 facing up.
   The die could land on a five 0, 1, 2, 3, or 4 times. So, f has the possible values of 0, 1, 2, 3, and 4.

My Turn!

Let b represent the number of boys in a family of 5. What are the possible values for b?

Objective 2: Classify data as a type of number.

You will encounter several classifications for data in the text. Although we will not go over those here, we will indirectly practice classifying, using some concepts you probably know from prerequisite material.

Example 3 Select which of the following types of numbers the number 6 belongs to.

Whole numbers, integers, real numbers

Whole numbers = {0, 1, 2, 3, 4, ...}

Integers = {..., −3, −2, −1, 0, 1, 2, 3,...}

Real numbers = all the numbers on a number line; this includes rational and irrational numbers.

The number 6 belongs to each of these sets of numbers.

My Turn!

Select which of the following types of numbers the number 2.5 belongs to.

Whole numbers, integers, real numbers

Objective 3: Round decimals.

Throughout the text, you will be asked to round your final answer.

Let's begin with a refresher on place value.

The table below indicates the place value for each digit in the number 1234.5678.

1	2	3	4	.	5	6	7	8
Thousand s	Hundred s	Ten s	One s	Decima l Point	Tenth s	Hundredth s	Thousandth s	Ten- thousandth s

You will often be asked to round your final answer to a particular place value. So, we will review the rounding basics in this first example.

Example 4 Round 3456.789 inches to the nearest tenth of an inch.

Step 1: Locate the digit in the desired place value. Let's call this digit R for purposes of our discussion.

The digit 7 is in the tenth place.

We are trying to decide whether the given number is closer to 3456.7 or 3456.8

Step 2: If the digit in the place value immediately to the right of R is a 0, 1, 2, 3, or 4, then keep R as it is and all digits to the right of R are zeros. If the zeros are not placeholders (that is, removing them does not result in other digits losing their place value), then they can be removed. If the digit in the place value immediately to the right of R is a 5, 6, 7, 8, or 9, then increase R by 1 (round up) and all digits to the right of R are zeros. Again, if these zeros are not placeholders, then they can be removed.

Since the digit immediately to the right of 7 is an 8, we must round up to 3,456.800. As the zeros are not placeholders, we can remove them.

My Turn!

Round 7654.32198 miles to the nearest hundredth of a mile.

Example 5 Round 0.050748 cm to the nearest ten-thousandth.

By rounding to the nearest ten thousandth, we are trying to decide whether 0.050748 is closer to 0.0507 or 0.0508. Since the digit immediately to the right of 7 is 4, we will keep the third significant digit at 7.

My Turn!

Round 0.36409 cm to nearest thousandth.

Objective 4: Convert between decimals, fractions, and percents.

There will be many times in this course that you will want to convert between decimals, fractions, and percents. We will review how to do this for the most commonly used conversions in statistics.

Example 6 Convert 8% to a decimal.

To convert from a percentage to a decimal, you will remove the percent symbol and move the decimal point two places to the left. (This is because the percent symbol means per hundred; so, you are essentially dividing by 100.)

Currently, the decimal point is to the right of 8. (An invisible decimal point is always to the right of the rightmost digit.) So, we will move it two places to the left. Since there is only one digit to the left, we will have to insert a placeholder of 0. Also, it is good practice to write a 0 to the left of the decimal point, as it results in less confusion to the reader. (Decimal points are pretty tiny for reading.)

$$8\% = 0.08$$

My Turn!

Convert 0.3% to a decimal.

Example 7 Convert 0.006 to a percent.

Now, we will do the reverse and convert a decimal to a percent. To do this you will move the decimal point two places to the right and insert the percent symbol.

$$0.006 = 0.6\%$$

My Turn!
Convert 2.3 to a percent.

Example 8 Convert $\frac{7}{8}$ to a percent.

First, we will convert the fraction to a decimal. Then, we will follow the steps from Example 7 to change the decimal to a percentage.

To change a fraction to a decimal, you divide the numerator by the denominator.

$$\frac{7}{8} = 7 \div 8 = 0.875$$

Now, we will convert the decimal 0.875 to a percentage by moving the decimal point two places to the right and inserting the percent symbol.

$$0.875 = 87.5\%$$

My Turn!

Convert $\frac{9}{16}$ to a percent.

Example 9 Convert 0.06 to a fraction.

To convert a decimal to a fraction, you write the number with the decimal point and leading zeros removed as the numerator of a fraction over a denominator that is a power of 10. The exponent on the 10 is the number of digits to the right of the decimal point in the given number. For 0.06, there are 2 digits to the right of the decimal point. So, the denominator will be 10^2, which can be rewritten as 100. Once you have a fraction, be sure to simplify it fully.

$$0.06 = \frac{6}{10^2} = \frac{6}{100} = \frac{3}{50}$$

My Turn!

Convert 0.0028 to a fraction.

Objective 5: Calculate relative frequencies.

As you construct tables and graphs in the text, you will be calculating relative frequencies, which are the ratio of the amount of times an event occurs to the number of times the event could occur. In other words, it is the number of times something happens divided by the total number of times it could happen. Relative frequencies are often represented as percentages.

Example 10 If 30 out of 80 people responded that they skip flossing at least one day a week, calculate the percentage of respondents that skip flossing at least one day a week.

So, the relative frequency for this problem is 30 out of 80, which can be written as $\frac{30}{80}$. We can then turn this fraction into a decimal. Finally, we convert our answer to a percentage by moving the decimal point two places to the right and inserting %.

$$\frac{30}{80} = 30 \div 80 = 0.375 = 37.5\%$$

My Turn!

If 450 out of 800 people responded that they own more than one computer, calculate the percentage of respondents who own more than one computer.

Objective 6: Find the percentage of a number.

Sometimes, you may need to determine what the percentage of a number is.

What is $p\%$ of x?

To do this, you will multiply the percentage written as a decimal times the number.

That is,

$$p\% \text{ as a decimal} \cdot x$$

Example 11 What is 30% of 150?

Write 30% as a decimal and multiply it by 150.

$$30\% \text{ of } 150 = 0.3 \cdot 150 = 45$$

My Turn!

What is 65% of 600?

Example 12 There were 900 respondents to a survey. If 40% of respondents indicated that they live in an apartment, how many respondents live in an apartment?

This amounts to answering the question, "What is 40% of 900?"

To find the percentage of a number, you multiply the percent written as a decimal times the number.

$$x = 0.4 \cdot 900 = 360$$

My Turn!

There were 80 respondents to a survey. If 30% of respondents indicated that they own a DVD player, how many respondents own a DVD player?

Chapter 2 Picturing Variation with Graphs

Objective 1: Position values on a number line.

Within the textbook, you will be organizing data into various graphs. These graphs will require that you are familiar with the placement of values on the number line. Let's take a moment to review the basics of the number line. The real number line is centered at 0 and extends in the negative direction to the left and to positive values on the right. Here is a common representation of the real number line.

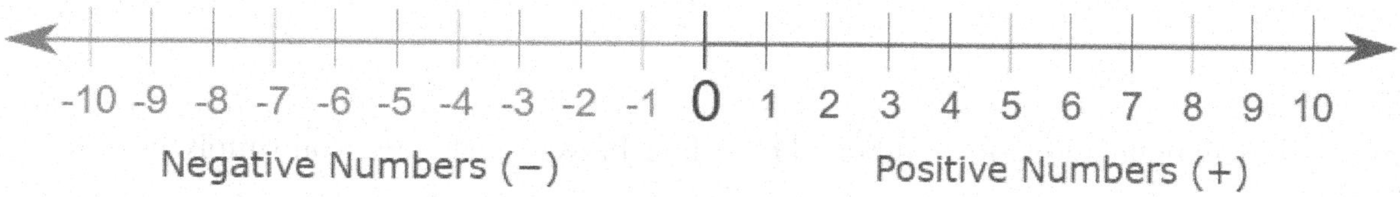

Most of the time, we deal with non-negative numbers when studying data. Therefore, we will focus on that portion of the number line.

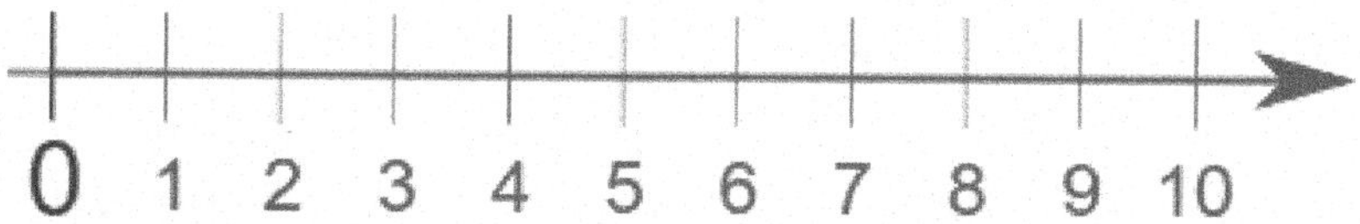

Example 1 Represent the following data on a number line.

The age at which a sample of kids first learned to ride a bike

4 years old, 5 years old, 7 years old

We will represent each data value with a dot in the appropriate location on a number line.

My Turn!

Represent the following data on a number line.

The number of keys on a keychain

1 key, 3 keys, 9 keys

Of course, not all data are whole numbers. For values between integers, you simply have to eyeball it.

Example 2 Represent the following data on a number line.

The following are the lengths (in centimeters) of the thumbs for a sample of four people.

Person 1: 4.5 cm

Person 2: 5.1 cm

Person 3: $5\frac{3}{4}$ cm

Person 4: $6\frac{7}{8}$ cm

First, you have to determine between which two integer values each data value lies.

4.5 lies midway between 4 and 5.

5.1 is just a little bit to the right of 5.0 on the number line, between 5 and 6.

For person 3, you may want to change the fraction to a decimal to assist you in visualizing its location. Many students find that they are more comfortable working with decimals than fractions, especially when the fractions become more complicated. $5\frac{3}{4}$ is equal to 5.75. Therefore, this value will be located between 5 and 6, and will be closer to 6.

Similarly, $6\frac{7}{8}$ is 6.875. This is between 6 and 7, just a tiny bit to the left of 7.

Here are the data values represented as points.

My Turn!

Represent the following data on a number line.

The following are the lengths (in inches) of the hairs cut off in a haircut for a sample of three hairs.

Hair 1: $2\frac{3}{4}$ in

Hair 2: $4\frac{7}{16}$ in

Hair 3: $3\frac{3}{8}$ in

Objective 2: Tally values.

When you are visualizing data, you often have to tally the amount of values that fit the conditions of a particular category.

Example 3

The following list of values represents the weights (in ounces) of boxes of chocolate chip cookies on a grocery store shelf: 16.0, 16.1, 16.2, 19.8, 21.7, 20.9, 18.8, 18.3, 18.9, 17.6, 17.9, 20.7. Create a tally table using the following weight groupings: 16.0 – 16.9, 17.0 - 17.9, 18.0-18.9, 19.0 - 19.9, 20.0 - 20.9, 21.0 - 21.9.

We are essentially being asked to count the number of pieces of data that fit within the conditions of each weight grouping.

For the first group 16.0 – 16.9, there are three data values that fit within these boundary values, 16.0, 16.1, and 16.2. Therefore, the tally for this grouping is 3. The related row in a tally table would look something like the following:

Weight of Cookies(oz)	**Tally (or Frequency)**
16.0 - 16.9	III or 2

We need to conduct similar counts for the other categories. For the group 17.0 – 17.9, there are two data values that fit within these boundary values, 17.6 and 17.9. There are three values, 18.8, 18.3, and 18.9 that fit within the next grouping of 18.0 – 18.9. There is only one value, 19.8, in the grouping of 19.0 – 19.9. Next, the values of 20.9 and 20.7 fit within the category of 20.0 – 20.9. Finally, 21.7 is the only value that fits in the final category of 21.0 – 21.9.

With the counts tallied, we can now organize this information into a table.

My Turn!

The following list of values represent the weights (in ounces) of newborn kittens: 3.52, 3.48, 3.25, 3.89, 3.67, 3.15, 3.97, 3.76, 3,84, 3.34. Create a tally table for the following weight groupings: 3.00 – 3.24, 3.25 – 3.49, 3.50 – 3.74, and 3.75 – 3.99

Objective 3: Find proportion.

In the corresponding chapter of the text, you will begin interpreting graphs. The graphs may include some that you may have seen before in the media, such as bar charts, pie charts, and time series graphs. We will do a sampling of two here. Notice how the graphs that follow are carefully labeled so that the information is clear to the reader. Whenever you create your own graphical representations of data, label everything clearly so that the graph conveys all the necessary information to the reader. We will practice finding a desired proportion using the information from a graph.

Example 4 The following is called a bar chart. It represents the types of patients that a veterinarian had in one day. What proportion of the vet's patients that day were cats? Round the answer to the nearest hundredth.

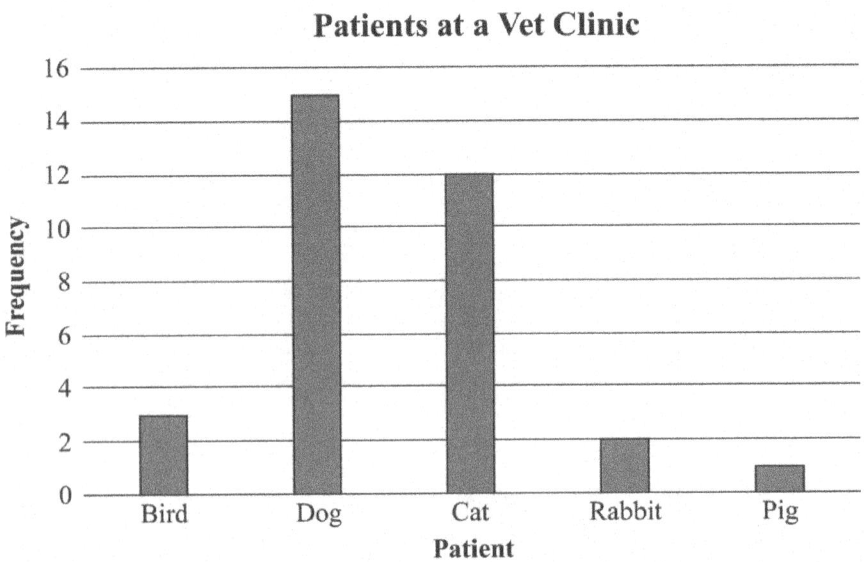

In order to answer this, we need to figure out how many cats were patients that day and how many animals were seen that day at the veterinarian's clinic.

The height of the bar labeled "Cat" will tell us that there were 12 cats seen that day.

To find the total number of animals seen that day, we have to sum up all of the heights (frequencies) of the bars.

Total number of animals seen that day $=3+15+12+2+1=33$.

So, to find the proportion, we have to divide 12 by 33 to change it to a decimal: $\frac{12}{33}=0.\overline{36}$. (The 36 has a bar above it since those digits are repeating.) We will now round to the nearest hundredth.

$$0.\overline{36}=0.3636\ldots\approx 0.36$$

My Turn!

The bar chart represents the frequencies for the types of rooms that are in a home. What percentage of the house's rooms are bathrooms? Round the answer to the nearest hundredth.

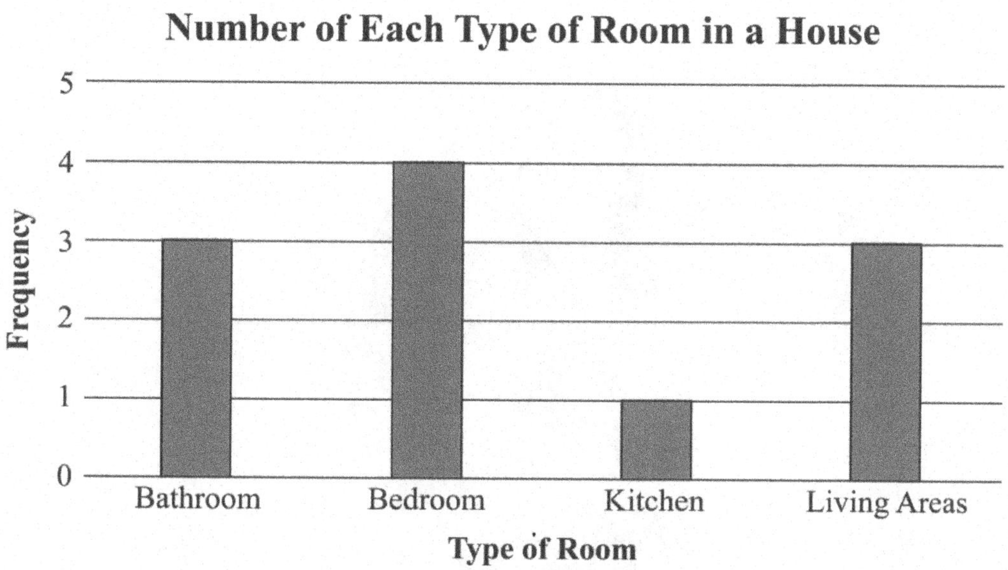

Example 5 The following is called a pie chart. It represents the college majors for a sample of students. What proportion of the college students were math majors? Round the answer to the nearest hundredth.

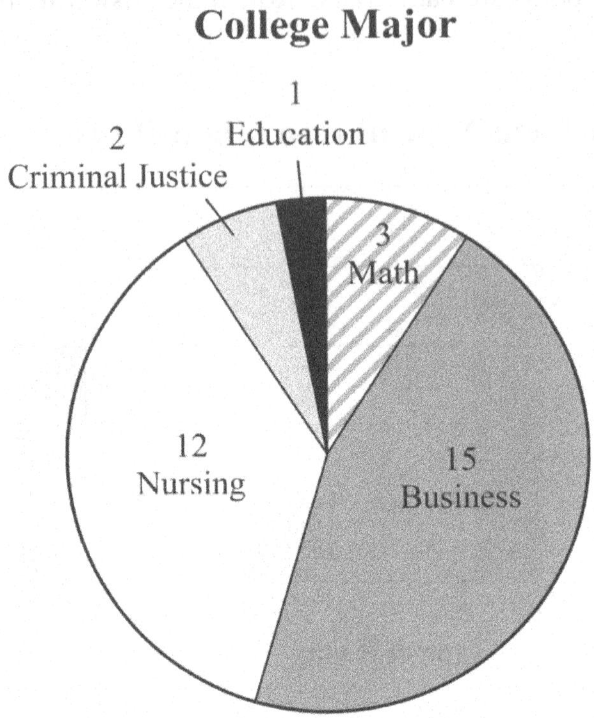

First, we need to figure out how many students have math as their college major. There were 3 math majors, as indicated by the number 3 in the appropriate slice of the pie chart. Next, we need to figure out the total number of students who were included in the sample. We can do that by summing up the frequencies for the 5 college majors.

Total number of students sampled $= 3 + 15 + 12 + 2 + 1 = 33$.

So in order to find the relative frequency, we have to divide 3 by 33 to change it to a decimal: $\frac{3}{33} = 0.\overline{09}$.

We will now round to the nearest hundredth.

$$0.\overline{09} \approx 0.090 \approx 0.09$$

My Turn!

The pie chart represents the frequencies for various types of toys in a pediatrician's waiting room. What proportion of the toys were dolls? Round the answer to the nearest hundredth.

Toys in a Pediatricians's Waiting Room

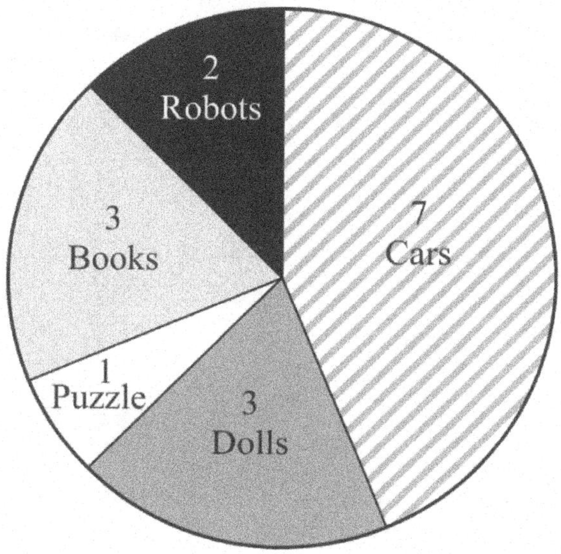

Chapter 3 Numerical Summaries of Center and Variation

Objective 1: Evaluate squares.

Let us review the meaning of squaring a number. For any real number x, $x^2 = x \cdot x$. That is, for any real number, to "square" a number means to multiply a number by itself.

Therefore, $3^2 = 9$ since $3^2 = 3 \cdot 3 = 9$.

Note that the number 3 is called the "base" and that the superscript of 2 is called the "exponent".

Example 1 Evaluate 15^2.

The exponent of 2 is telling us to multiply the base number 15 by itself.

$$15^2 = 15 \cdot 15 = 225$$

My Turn!

Evaluate 25^2

Objective 2: Evaluate square roots.

Recall the meaning of square root.

The principal square root of a number a is denoted by $\sqrt{a}$. Furthermore, $x = \sqrt{a}$ if $x^2 = a$. So, $3 = \sqrt{9}$, since $3^2 = 9$.

Note that for the order of operations a square root can be treated as an exponent.

Example 2 Evaluate $\sqrt{64}$.

We are essentially trying to find a number such that the number times itself is 64.

$$\sqrt{64} = \sqrt{8 \cdot 8} = 8$$

My Turn!

Evaluate $\sqrt{25}$.

If the value under the radical symbol (i.e. the radicand) is not a perfect square, you are going to have to try a different strategy. You may recall having to simplify radicals for algebra. However, for the way that you will be using square roots for statistics, you won't have to simplify square roots as you would in an algebra course. You will simply use a calculator for radicands that are not perfect squares. However, you may want to come up with an estimate of which two integers the square root is between in order to catch careless calculator entry errors.

Example 3 Which two integers is $\sqrt{88}$ between?

You need figure out which two perfect squares 88 is between.

$$81 < 88 < 100$$

$$9^2 < 88 < 10^2$$

So,

$$9 < \sqrt{88} < 10$$

My Turn!

Which two integers is $\sqrt{65}$ between?

Example 4 Approximate $\sqrt{88}$ to the nearest hundredth.

Use a calculator to evaluate this, since 88 is not a perfect square.

For the TI-83/84 calculators, the keystrokes are 2nd $\sqrt{\ }$ 88 enter. Note that the $\sqrt{\ }$ is the second function for the x^2 button. If you need to find other roots, press MATH.

Your calculator output is approximately 9.3808. This rounds to 9.38, which agrees with our findings from Example 5, as it is between 9 and 10.

My Turn!

Approximate $\sqrt{130}$ to the nearest hundredth.

Objective 3: Apply the order of operations.

We use the order of operations to simplify a numerical expression. You will have many numerical expressions that you will have to simplify in statistics, especially if you are evaluating formulas.

It is important that we do the operations in the correct order to arrive at a correct answer. It is critical that each step is completed from left to right.

Below you will find a listing of the order of operations. Always, calculate the operations from left to right. This is especially important to remember for steps 3 and 4. This means that if there is a division to the left of multiplication in an expression, you would perform the division first.

1. **P**arentheses

2. **E**xponents

3. **M**ultiply or **D**ivide

4. **A**dd or **S**ubtract

Example 5 Perform the indicated operations: $5 \cdot 8^3$.

First we must calculate the exponent, since exponents should be done before multiplication according to the order of operations. (We reviewed calculating exponents in Integrated Review Chapter 1.)	$5 \cdot \boxed{8^3} = 5 \cdot \boxed{512}$
Next, we multiply.	$5 \cdot 8^3 = 5 \boxed{\cdot} 512 = 2560$

My Turn!

Perform the indicated operations: $6^4 \div 4$.

Example 6 Perform the indicated operations: $15 \div 3 \cdot 2 + 9$.

First, we must multiply or divide from left to right. Since the division is to the left of the multiplication, we must do that first. It is a common misconception to think that you always multiply before dividing. This is not the case; you complete the operations of multiplication and division as they appear from left to right.	$15\boxed{\div}3 \cdot 2 + 9 = \boxed{5} \cdot 2 + 9$
Now, we multiply.	$5\boxed{\cdot}2 + 9 = \boxed{10} + 9$
Finally, we perform addition.	$10\boxed{+}9 = 19$

My Turn!

Perform the indicated operations.

$5 + 2 \cdot 8$

Example 7: Perform the indicated operations: $\dfrac{(5-3)^2 + (7-3)^2 + (9-3)^2}{7}$

Round to the nearest tenth, as needed.

First, we must simplify within each set of parentheses.	$\dfrac{(5-3)^2 + (7-3)^2 + (9-3)^2}{7} = \dfrac{2^2 + 4^2 + 6^2}{7}$
Now, calculate the exponents.	$= \dfrac{4 + 16 + 36}{7}$

You can think of the entire numerator as being within an invisible set of parentheses. So, you would simplify that next.	$=\frac{56}{7}$
Finally, you can perform the division (or reduce the fraction).	$=8$

My Turn!

Perform the indicated operations: $\frac{10}{(5+3)^2+(8+3)^2}$.

Round to the nearest tenth, as needed.

Objective 4: Evaluate expressions and formulas.

Example 8 Perform the indicated operations:

$$\frac{18-11}{\frac{3}{\sqrt{9}}}$$

In order to simplify this expression, we will apply the order of operations to an expression that contains a square root. We treat a radical symbol as an exponent in the order of operations. We can do this because radicals can be rewritten as rational exponents (e.g. $\sqrt{9}=9^{\frac{1}{2}}$).

We can begin by viewing the numerator 18 – 11 as being contained within invisible parentheses and the denominator $\frac{3}{\sqrt{9}}$ as being contained within another set of parentheses. We can work on simplifying each of these. The simplification of the numerator is shown first.	$\frac{(18-11)}{\left(\frac{3}{\sqrt{9}}\right)}=\frac{7}{\left(\frac{3}{\sqrt{9}}\right)}$
To simplify the denominator $\frac{3}{\sqrt{9}}$, we must find $\sqrt{9}$.	$\frac{7}{\left(\frac{3}{\sqrt{9}}\right)}=\frac{7}{\left(\frac{3}{3}\right)}=\frac{7}{1}$
We find our final answer by dividing (or simplifying the fraction).	$\frac{7}{1}=7$

My Turn!

Perform the indicated operations:

$$\frac{6.8-6.1}{\frac{1.1}{\sqrt{16}}}.$$

Round to the nearest hundredth, as needed.

A formula is a special type of equation that expresses the relationship between several variables. On one side of the equation (usually the left side), you have the dependent variable. On the other side you have the independent variables. To evaluate a formula, we substitute known values for the dependent variables and follow the order of operations to simplify. We will include one example here, and you will see numerous examples in the chapters that follow.

Example 9 Evaluate the formula

$$m = t \cdot \frac{s}{\sqrt{n}}$$

for $t = 1.96$, $s = 2.1$, and $n = 100$. Round the answer to the nearest thousandth.

First, we must substitute the given values for the independent variables.

$$m = t \cdot \frac{s}{\sqrt{n}} = 1.96 \cdot \frac{2.1}{\sqrt{100}}$$

In order to simplify the right-hand side, we will apply the order of operations. to an expression that contains a square root.

We treat a radical symbol as an exponent in the order of operations. So, we will begin there. The square root of 100 is 10.	$m = 1.96 \cdot \frac{2.1}{\sqrt{100}} = 1.96 \cdot \frac{2.1}{10}$
We can simplify the fraction by dividing by 10.	$m = 1.96 \cdot \frac{2.1}{10} = 1.96 \cdot 0.21$

We find our value for the dependent variable, *m*, by multiplying.	$m = 1.96 \cdot 0.210 = 0.4116$
Now, we round our final answer to the nearest thousandth. Since the value in the ten thousandth place is a 6, we round to 0.412.	$m \approx 0.412$

My Turn!

Evaluate the formula

$$n = \frac{z^2 \cdot \hat{p} \cdot (1 - \hat{p})}{m^2}$$

For $z = 2.575$, $\hat{p} = 0.3$, and $m = 0.05$. Round the answer to the nearest whole number.

Objective 5: Apply summation notation.

You will often see the summation symbol, Σ, in statistical formulas. It is the capital Greek letter "sigma." Whenever you see this symbol in a formula, it is telling you to add up values.

For instance, $\bar{x} = \frac{\Sigma x}{n}$ is the formula for finding the mean (average), $\bar{x}$, where x represents the data values and n is the number of data values.

Example 10 Find $\bar{x} = \frac{\Sigma x}{n}$ for the values in the accompanying table.

x
34
48
25
76
22
29

There are 6 values for x, so $n = 6$.

The formula is telling us to add up the values of x and divide by n.

$$\bar{x} = \frac{\Sigma x}{n} = \frac{34+48+25+76+22+29}{6} = \frac{234}{6} = 39$$

Note that the above formula could also be written as

$$\sum_{i=1}^{6} \frac{x_i}{n}$$

My Turn!

Find $\bar{x} = \frac{\Sigma x}{n}$ for the values in the accompanying table, where n is the number of given values. Round to the nearest tenth, as needed.

x
5
4
8
7
2

Example 11 Find

$$\frac{\Sigma(x-39)^2}{n-1}$$

for the accompanying table and $n = 6$.

x
34
48
25
76
22
29

This notation is telling us to subtract 39 from each x value first, since we work out parentheses first in the order of operations. Next we would calculate exponents per the order of operations. Then, we sum up these squares of these differences. Once we have this sum, then we can perform division by the denominator.

$$\frac{\Sigma(x-39)^2}{n-1} = \frac{(34-39)^2+(48-39)^2+(25-39)^2+(76-39)^2+(22-39)^2+(29-39)^2}{6-1}$$

$$= \frac{(-5)^2+(9)^2+(-14)^2+(37)^2+(-17)^2+(-10)^2}{6-1}$$

$$= \frac{25+81+196+1369+289+100}{5}$$

$$= \frac{2060}{5}$$

$$= 412$$

My Turn!

Find

$\frac{\Sigma(x-2)^2}{n-1}$ for the accompanying table and $n = 4$. Round to the nearest tenth, as needed.

x
4
3
2
1

Chapter 4 Regression Analysis: Exploring Associations between Variables

Objective 1: Plot points.

In statistics, we will often study data that come in pairs. It can be helpful to see the paired data on a graph. Each set of paired data can be represented by a point that can be written as (x, y). The points can be plotted on a Cartesian coordinate plane. The origin, or the place where the horizontal (x) and vertical (y) axes intersect, is the starting point The first value in the paired data (the x value) will tell you to move left (negative) or right (positive). The second value (the y value) will place the point up (positive) or down (negative). For example, you can plot the set of paired data (3, 5) by counting three units to the right of the origin and up five units.

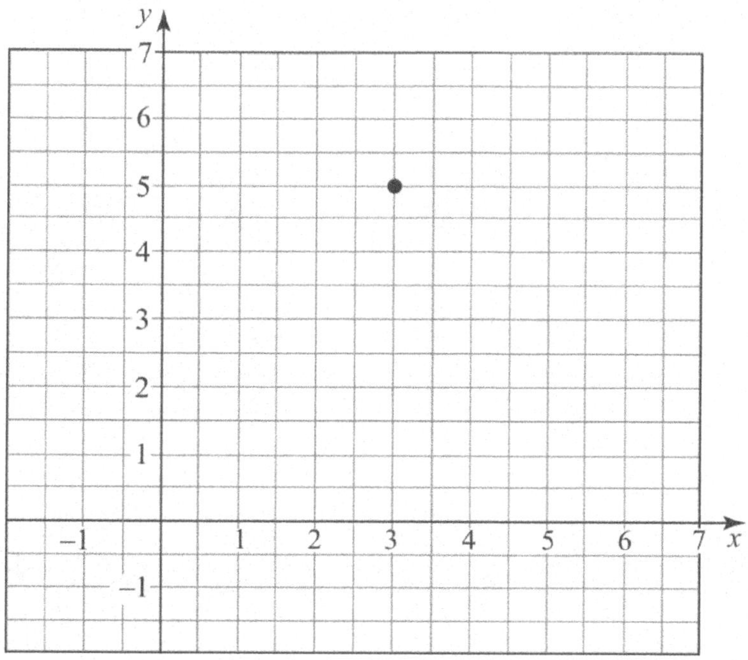

Example 1 Plot the following paired data.

Number of People in a Household (x)	Number of TVs in a Household (y)
1	2
2	4
3	5

These data can be written as three points (1, 2), (2, 4), and (3, 5). Then, we can plot these three points on the Cartesian plane. As you learn statistics, you will find that most of the data that you will encounter will be positive values. As a result, many times you can limit your Cartesian plane to the first quadrant (the upper right-hand section). Also, there may be times that you will want to change the scale on the axes. Be sure to label your axes with the variables that are being used.

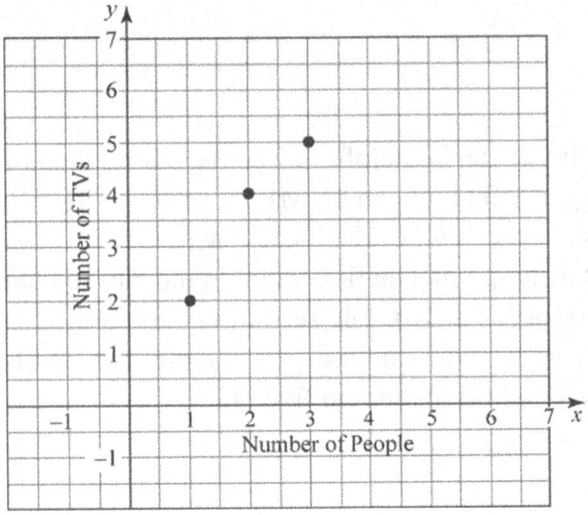

My Turn!

Plot the following paired data.

Number of Rooms in a House	Number of Paintings in a House
5	15
7	20
8	26

Objective 2: Review the skills for graphing.

We have already reviewed how to plot paired data as points on the Cartesian coordinate plane. You will often want to determine whether these points follow some sort of visible pattern. One common pattern to look for is whether the points fall on a line; that is, could you connect the points using a straightedge such as a ruler.

Example 2 Plot the points (2,4), (3, 5), and (5,7) and determine whether they fall on a straight line.

First we, plot the points. Recall that the first value in the paired data (the x value) will tell you to move left (negative) or right (positive) and the second value (the y value) will place the point up (positive) or down (negative). For example, you can plot the set of paired data (2, 4) by counting two units to the right of the origin and up four units.

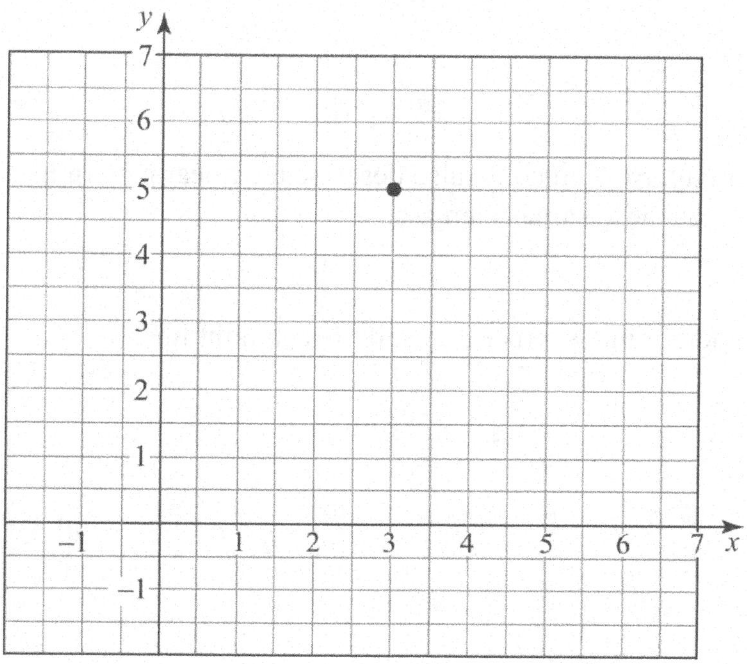

Once you have the plotted the data points, then you use your judgement to determine whether it seems reasonable to connect those points in a line. In this case, it appears that the points are indeed linear. Therefore, you can use a straightedge and connect these points in a line.

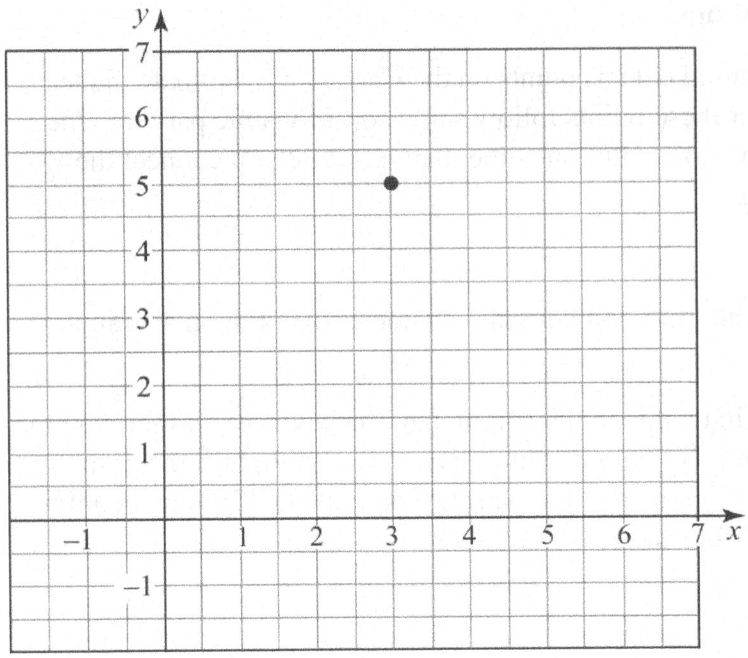

Yes, these points do indeed follow a linear pattern. You could also describe the linear pattern in more detail by saying that *y*-values increase as the *x*-values increase.

My Turn!

Plot the points (1, 5), (2, 4), and (3,3) and determine whether they fall on a straight line.

Example 3 Plot the following paired data. Does the data appear to be linear? Did the number of cell phones sold decrease, increase, or stay about the same? In which year were four million cell phones sold?

Number of Years Since 2000 (x)	**Number of Cell Phones Sold (in millions) (y)**
0	2
3	5
5	7

These data can be written as three points (0, 2), (3, 5), and (5, 7). Then, we can plot these three points on the Cartesian plane. In statistics, you will often adjust the scale so that the graphs will be more easily attainable. The data has already been given to us in scales that are easier to work with. Be sure to label your axes accordingly.

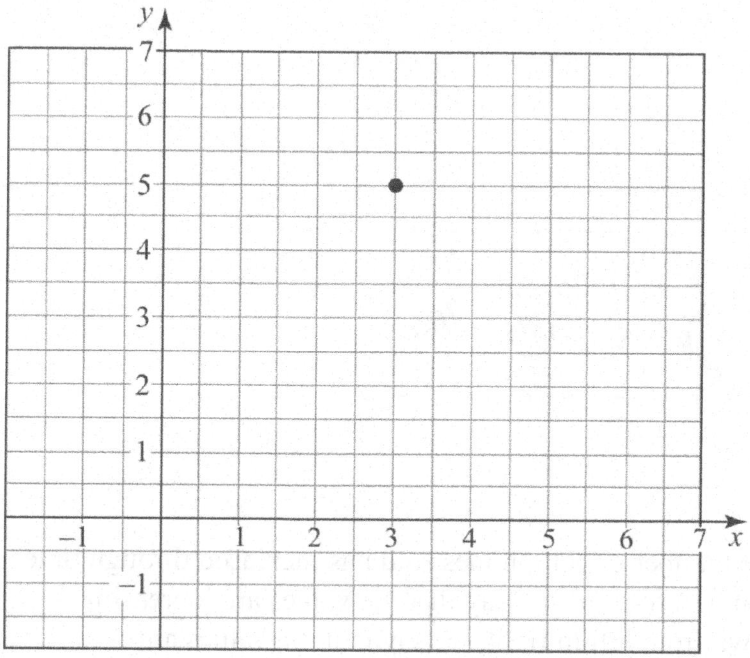

These points do indeed seem to follow a linear pattern.

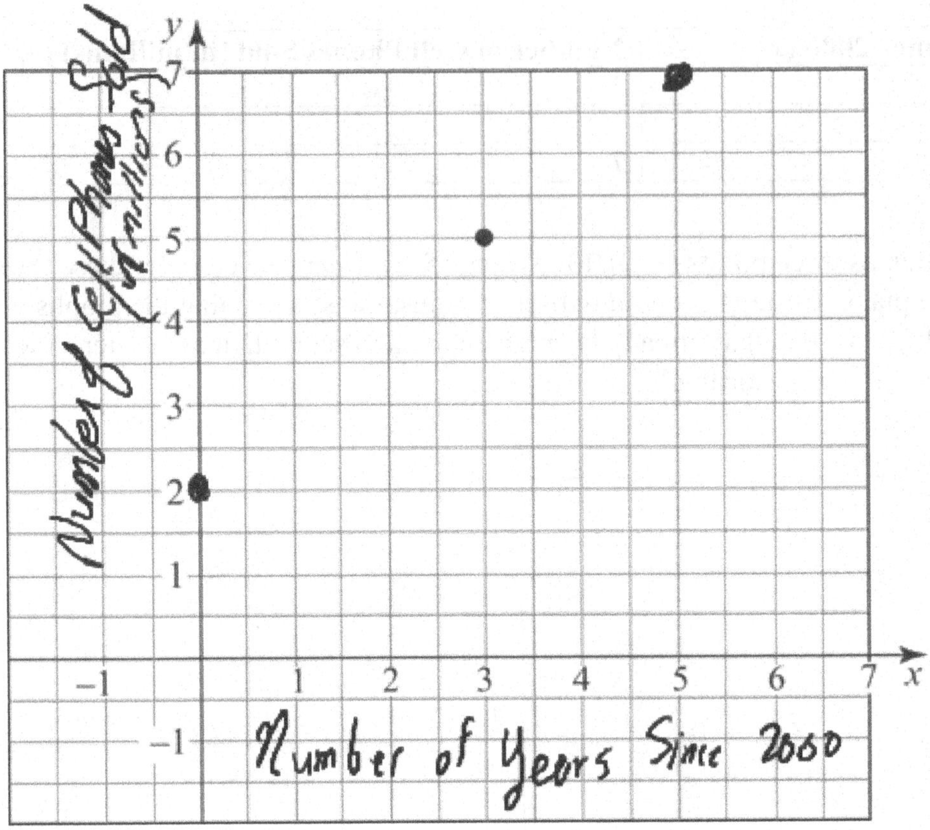

As the line goes up from left to right, the number of cell phones sold has increased through time. In general, when the line is going up from left to right, we say that the values are increasing through time. When the line is going down from left to right, we say that the values are decreasing through time. If the points are in a horizontal line, we say that the *y*-values are remaining constant through time.

Now, we need to find the year when four million cell phones were sold. So, we look at the *y*-axis for the value 4. Then we look to see where this hits the line we formed from the data. From this point of intersection, we travel down to the *x*-axis. The corresponding value on the *x*-axis is 2. This means that two years after 2000, which would be the year 2002, the company sold 4 million cell phones.

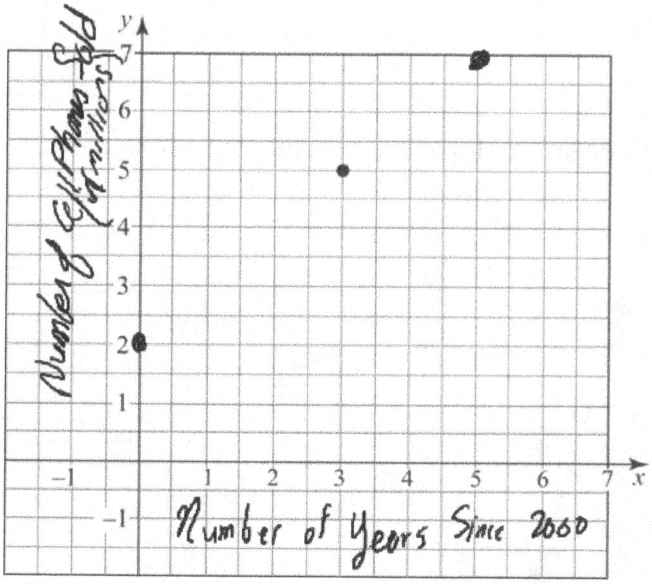

My Turn!

Plot the following paired data. Is the data linear? Has the average number of students in the classroom increased, decreased, or remained constant? In which year, was there an average of 23 students in the classroom?

Number of Years Since 2010	Average Number of Students in a Classroom
1	26
5	24
9	22

Objective 3: Find and interpret slope.

The slope of the line is its tilt or steepness.

Line with positive slope

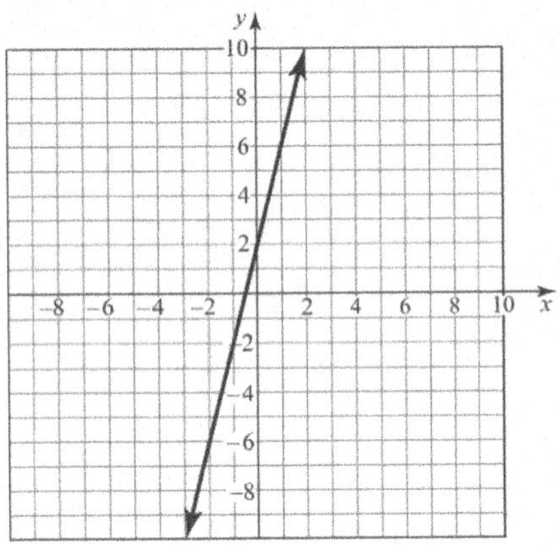

Line with negative slope

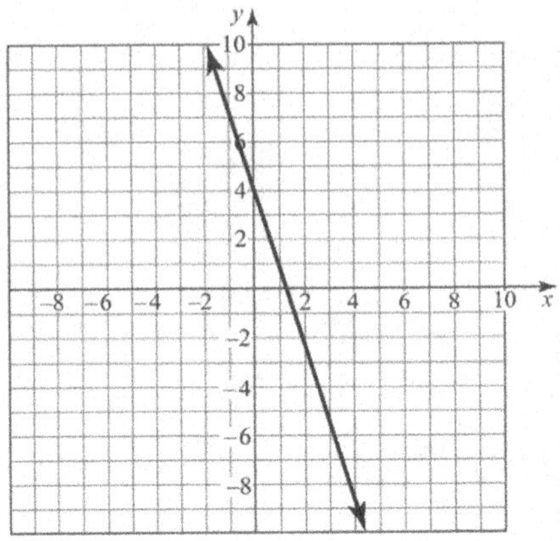

Line with slope of 0

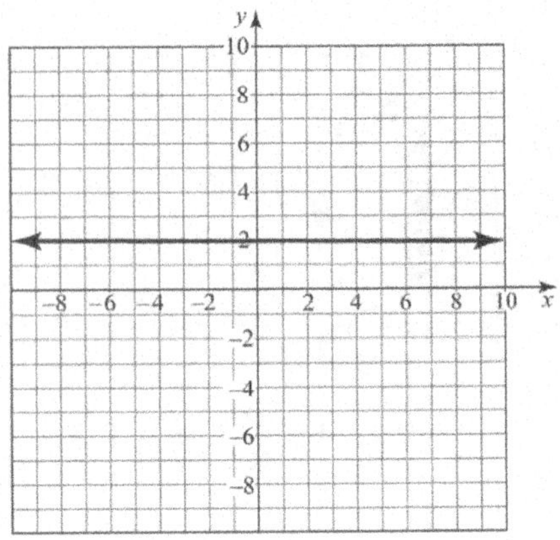

Here are some phrases related to slope that you may have heard:

$$Slope = \frac{rise}{run} = \frac{change\ in\ y}{change\ in\ x}$$

The slope of the line m between the points (x_1, y_1) and (x_2, y_2) can be given by

$$m = \frac{y_2 - y_1}{x_2 - x_1}$$

Example 4 Find the slope of the line that contains the points (3, 5) and (6, 9).

We can label the points as follows $(x_1, y_1) = (3,5)$ and $(x_2, y_2) = (6,9)$.

Now, we will substitute the given points into the formula for slope:

$$m = \frac{y_2 - y_1}{x_2 - x_1} = \frac{9-5}{6-3} = \frac{4}{3}$$

My Turn!

Find the slope of the line that contains the points (0.5, 2) and (0.7, 5). Round the answer to the nearest hundredth.

Example 5 Find the slope of the line given in the following graph.

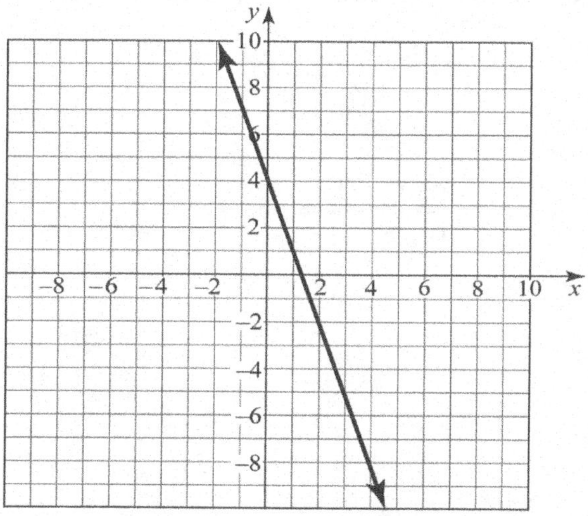

Begin by locating two points on the line. We will use the points $(1, 1)$ and $(2, -2)$, although you could use any two points on the line.

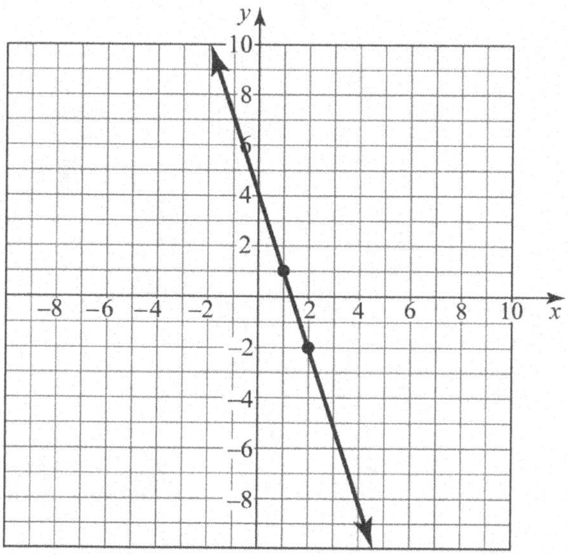

You could then either use the formula or map out the rise over run from the graph.

Formula Method: $m = \frac{y_2 - y_1}{x_2 - x_1} = \frac{-2-1}{2-1} = \frac{-3}{1} = -3$

Rise-Over-Run Method: Let's look at this problem as if we are travelling from $(2,-2)$ to $(1, 1)$. The rise from -2 to 1 is 3 units. The run from 2 to 1 is -1. The run is negative because we went to the left which is the direction of the negatives on the number line. (You could have also traveled from $(1, 1)$ to $(2,-2)$ which has a rise of -3 and a run of 1, resulting in the same slope.)

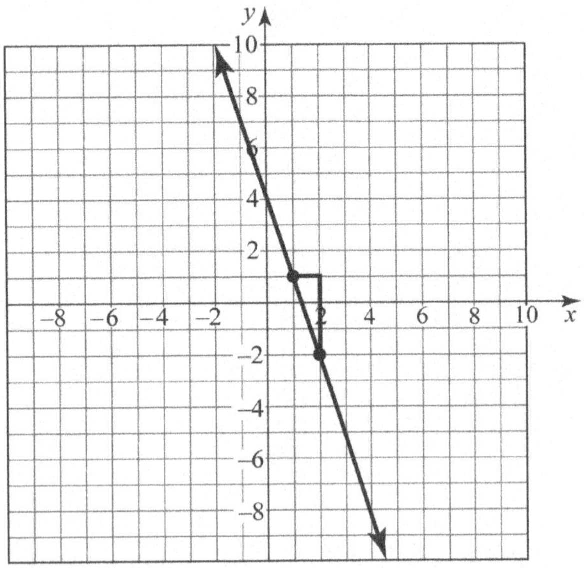

$m = \frac{\text{rise}}{\text{run}} = \frac{3}{-1} = -3$

My Turn!

Find the slope of the line given in the following graph.

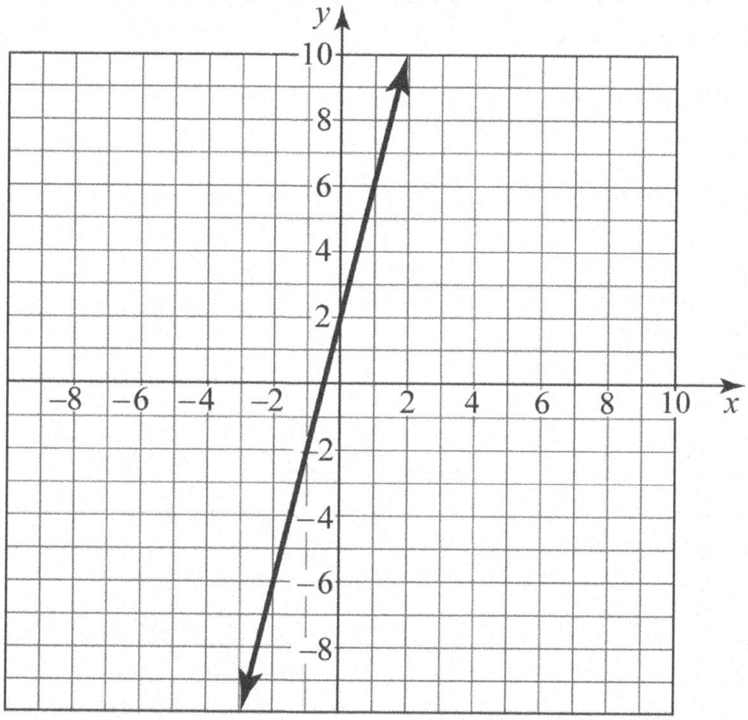

Objective 4: Find the *y*-intercept of a line.

The *y*-intercept(s) of a graph is (are) the point(s) where a graph crosses the *y*-axis. For a line, there is at most one *y*-intercept. Any *y*-intercept should be written as an ordered pair in the format (0, *b*), where *b* is the *y*-coordinate of the *y*-intercept.

Example 6 Find the *y*-intercept of the following graph.

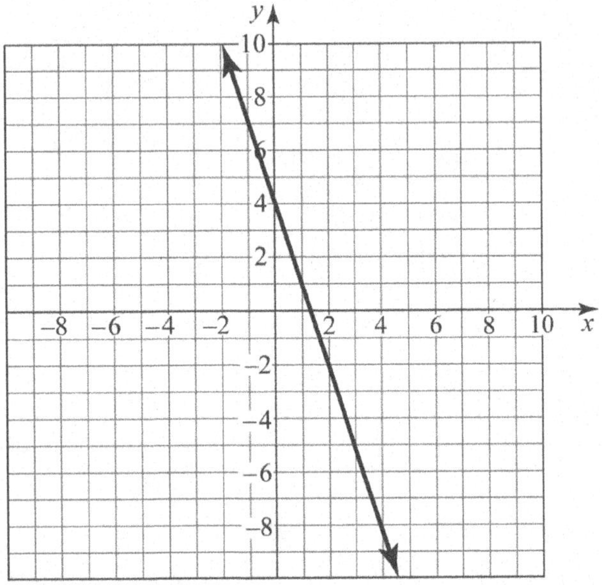

Looking at the graph, we can see that the line clearly crosses through the *y*-axis at the point (0,4).

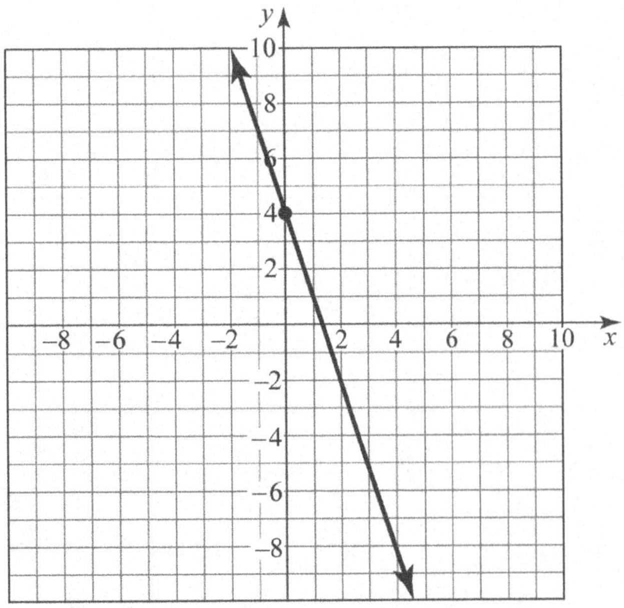

My Turn!

Find the y-intercept of the following graph.

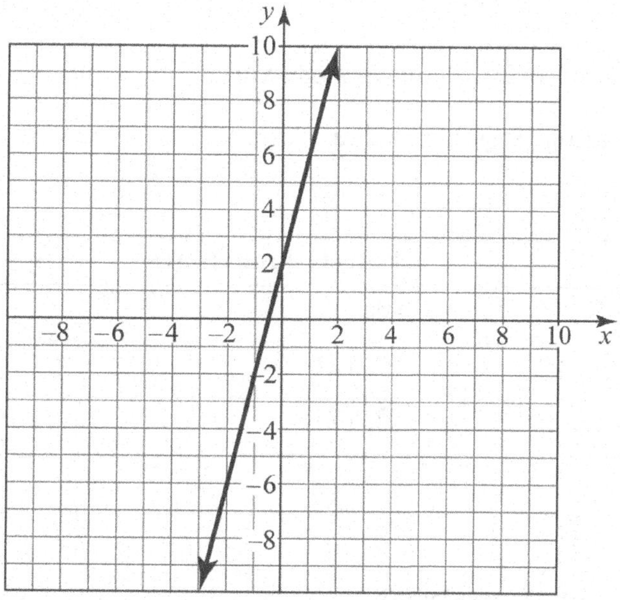

For any linear equation written in slope-intercept form, $y = mx + b$, the y-intercept can be written as $(0, b)$. That is, the constant value b is the y-coordinate of the y-intercept.

Example 7 Find the y-intercept of the line $y = 7x - 3$.

The equation is written in the form $y = mx + b$. The value of b is -3. Since the y-intercept should be written as an ordered pair $(0, b)$, the y-intercept for this line is $(0, -3)$.

My Turn!

Find the y-intercept of the line.

$\hat{y} = \frac{2}{3}x + \frac{1}{2}$

Objective 5: Find values from a linear equation or graph.

Example 8 Find three pairs of data values that would be on the line $\hat{y} = 5x + 7$.

There are actually an infinite number of points on any line. So, answers can vary.

If you plug in any *x* value, you will get a corresponding $\hat{y}$ (*y*-hat) value.

Let's plug in 0, 1, and 2 for *x*.

x	$\hat{y} = 5x + 7$	**Paired data values**
0	$\hat{y} = 5(0) + 7 = 0 + 7 = 7$	(0, 7)
1	$\hat{y} = 5(1) + 7 = 5 + 7 = 12$	(1, 12)
2	$\hat{y} = 5(2) + 7 = 10 + 7 = 17$	(2, 17)

My Turn!

Find three pairs of data values that would be on the line $\hat{y} = -2x + 3$.

Example 9 In the following graph, the horizontal axis represents the height of human beings and the vertical axis represents the weight of human beings. Find the weight (in pounds) for somebody who is 60 inches tall.

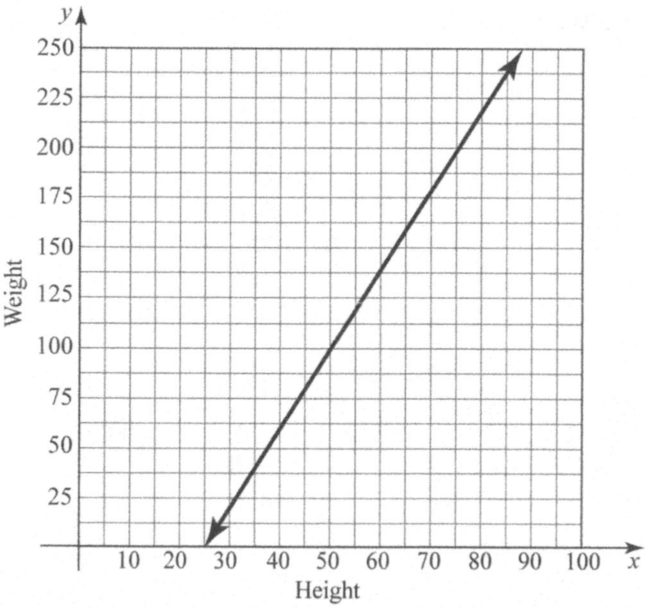

The horizontal axis represents height and the vertical axis represents weight. So, we begin by locating the given value of height = 60 on the horizontal axis and then move up to locate the point on the given line. Then, we look over to the vertical axis and find the corresponding *y* value.

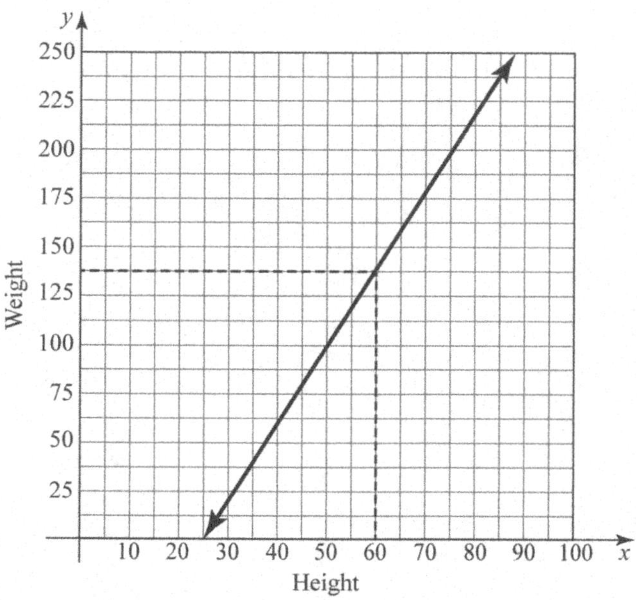

The corresponding weight for a height of 60 inches looks to be about 137.5 pounds, since it halfway between 125 and 150 pounds.

My Turn!

In the following graph, the horizontal axis represents the weight (in pounds) of dogs and the vertical axis represents the weight (in pounds) that the dog eats in a week. How many pounds of dog food does a dog who weighs 48 pounds eat in a week?

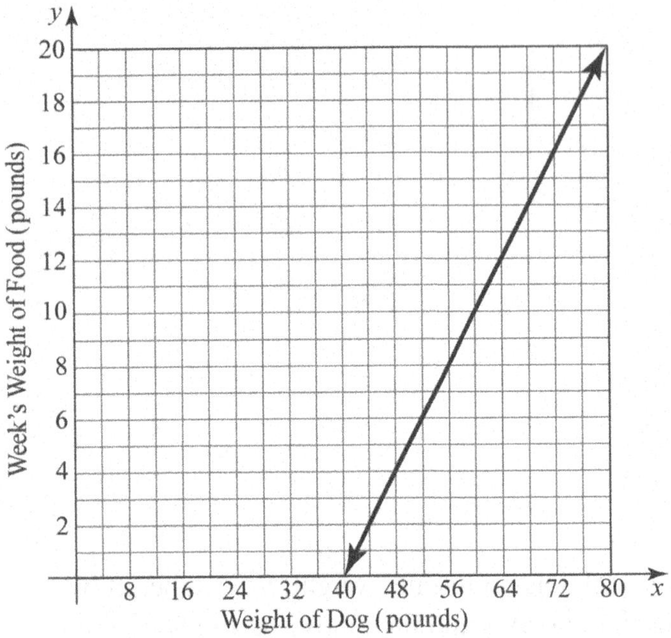

Objective 6: Graph a linear equation.

Example 10 Graph $y = -\frac{2}{3}x + 5$.

There are quite a few strategies for graphing lines from an equation. For this example, we will use the slope and the *y*-intercept to graph the line.

First, extract the *y*-intercept from the linear equation. Since $b = 5$, the *y*-intercept for this line is (0, 5). Plot the *y*-intercept.

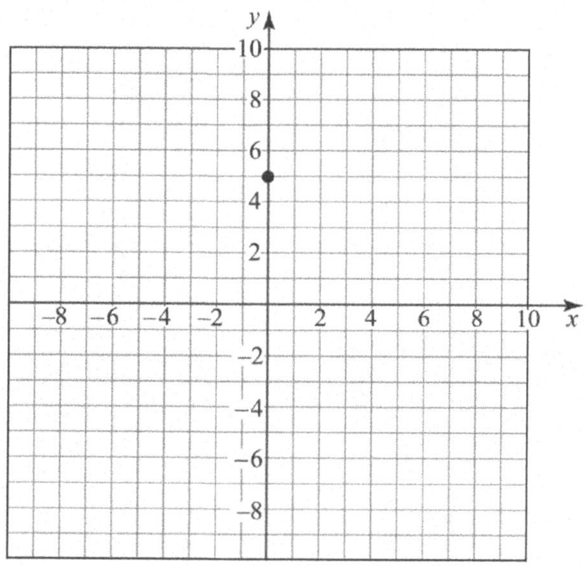

Once you have the *y*-intercept, you can use the slope to locate another point. The slope in this problem is $-\frac{2}{3}$. So, another point will be down 2 units and to the right 3 units. (Remember rise/run.)

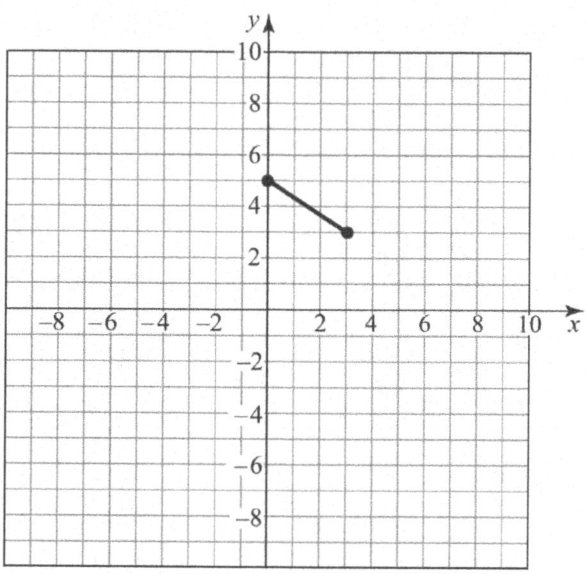

You can now extend this line using a straightedge.

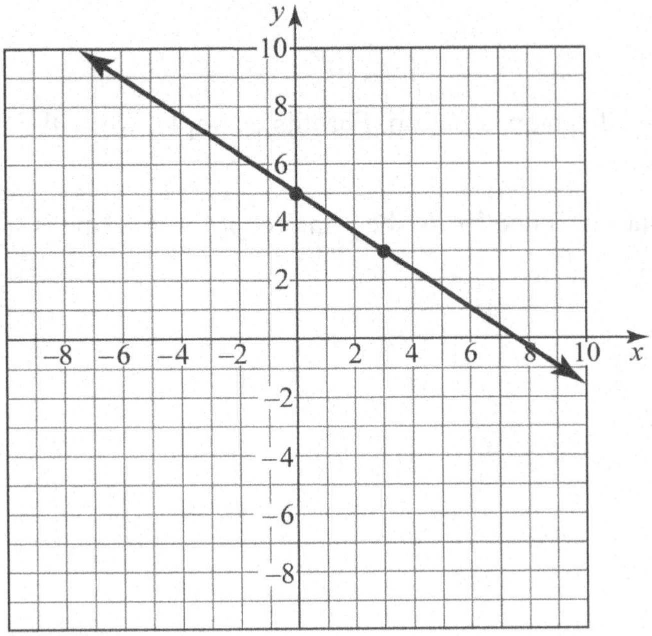

My Turn!

Graph $y = \frac{1}{4}x + 3$.

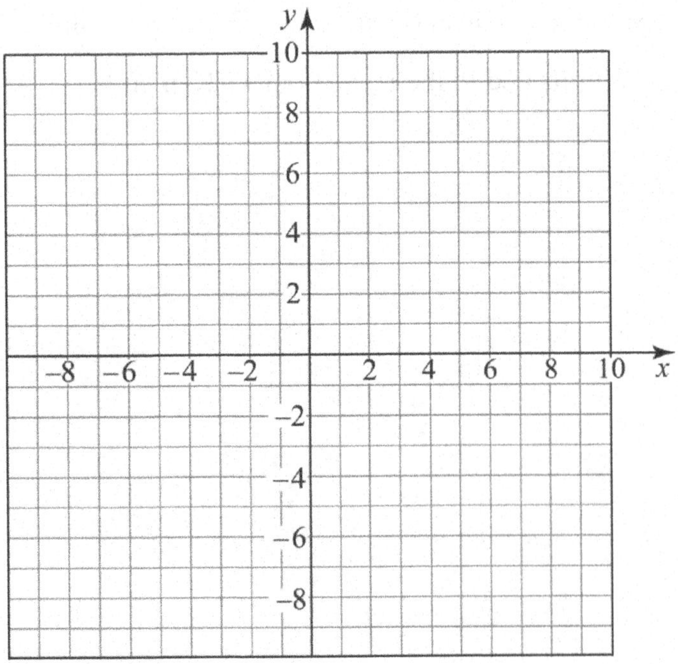

Example 11 Graph $y = 4x - 10$.

You could graph this equation using the method of Example 6 by using the y-intercept of $(0, -10)$ and the slope of 4. However, we will review another method in this example. You can find two points on the line and then connect them with a straightedge. Finding a third point is recommended as a check (or you could use the y-intercept as a check).

x	$y = 4x - 10$	**Paired data values**
1	$y = 4(1) - 10 = 4 - 10 = -6$	(1, −6)
2	$y = 4(2) - 10 = 8 - 10 = -2$	(2, −2)

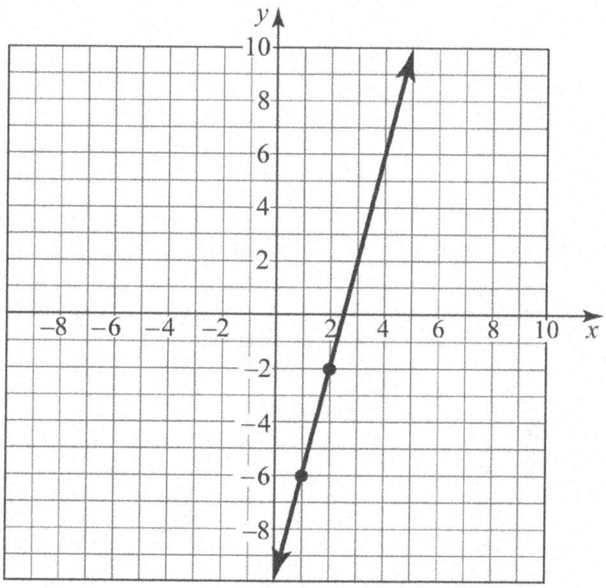

We can see that the line also passes through the y-intercept $(0, -10)$, which serves as our check.

My Turn!

Graph $y = \frac{1}{2}x - 5$. (Hint: You may want to substitute values for x that are multiples of the denominator.)

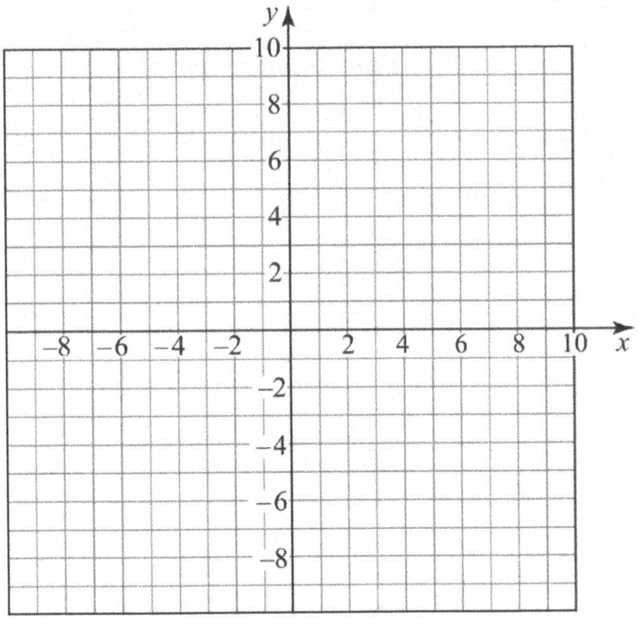

Objective 7: Find and interpret a linear model.

You will be studying linear models in great detail in the textbook. For those problems, you will be looking at real data that can sometimes be messy and may best be handled with a calculator. In this workbook, we will use a simplistic model to help prepare you.

Example 12 Find a linear model for the following data for the number of pages in a book and its weight in ounces.

Number of Pages	**Weight (oz.)**
150	6
200	8
250	10
300	12

The first thing you may want to do is plot the points to determine whether they fall on a line.

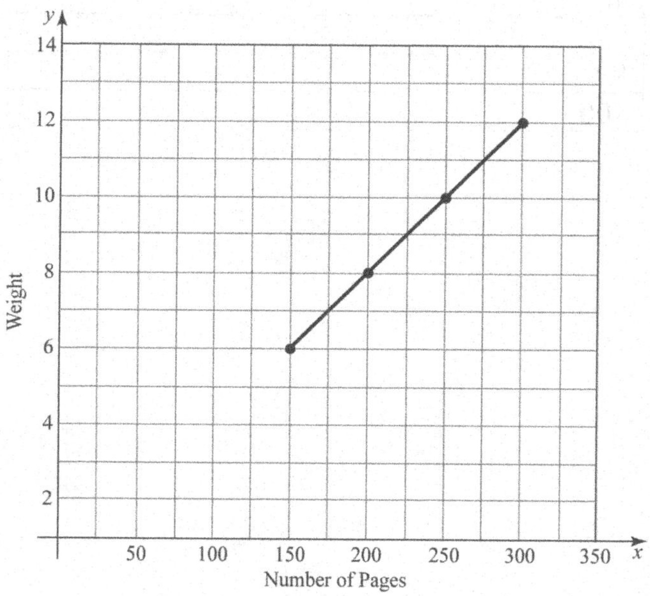

It does appear that the data fall on a line. So, we will use two points to find the slope.

$$m = \frac{8-6}{200-150} = \frac{2}{50} = \frac{1}{25}$$

We now need to find the y- intercept. Since it was not one of the given points, we can substitute any given point for x and $\hat{y}$ into $\hat{y} = \frac{1}{25}x + b$ and solve for b.

$$\hat{y} = \frac{1}{25}x + b$$
$$6 = \frac{1}{25}(150) + b$$
$$6 = 6 + b$$
$$0 = b$$

So, the linear model that fits the data is $\hat{y} = \frac{1}{25}x + 0$.

My Turn!

Find a linear model for the following data for the number of hours spent studying and the final exam score.

Number of Hours Spent Studying	Final Exam Score
4	40
6	50
12	80
18	100

Chapter 5 Modeling Variation with Probability

Objective 1: Evaluate exponents.

Recall that exponents are shorthand for repeated multiplication. The value of the exponent tells you how many times the factor repeats.

For instance, $3^4 = 3 \cdot 3 \cdot 3 \cdot 3$.

Example 1 Evaluate 8^3.

We expand 8^3, and then calculate the repeated multiplication. Most technology can calculate exponents, and you may want to explore how to do this on your calculator now.

$$8^3 = 8 \cdot 8 \cdot 8 = 512$$

My Turn!

Evaluate 7^4.

Objective 2: Write a fraction in lowest terms.

There are times that your result may be a fraction, especially when we work on the chapter on probability. Whenever you give a final result as a fraction, it must be fully reduced. In order for a fraction to be in lowest terms, there should be no common factors in the numerator and denominator.

Example 2 Write $\frac{48}{72}$ in lowest terms.

We may not all arrive at a reduced fraction by taking the same route. What is key is that we all check to make sure that there are no common factors left in the numerator and denominator. If there are any, cancel them out, since $\frac{a}{a} = 1$.

$$\frac{48}{92} = \frac{2 \cdot 24}{2 \cdot 46} = \frac{24}{46} = \frac{2 \cdot 12}{2 \cdot 23} = \frac{12}{23}$$

Note that many calculators (including the TI 83/84) can rewrite fractions in lowest form for you. You may want to become comfortable using this feature, especially for fractions with large numerators and denominators.

My Turn!

Write $\frac{36}{102}$ in lowest terms.

Objective 3: Apply operations to fractions.

We will review the basic operations of addition, subtraction, multiplication, division, and exponentiation of fractions. Although you may typically calculate these using a calculator within a statistics course, we will briefly review how to do them without technology. However, you may also want to try each operation using a calculator to make sure you know how to work with fractions on your calculator.

Adding Fractions

Example 3 Perform the indicated operation:

$$\frac{3}{16}+\frac{9}{16}$$

In order to add fractions, you need a common denominator. The two given fractions already have the common denominator of 16. So, all you need to do is add the two numerators together and keep the common denominator. Always write your answers in reduced form.

$$\frac{3}{16}+\frac{9}{16}=\frac{(3+9)}{16}=\frac{12}{16}=\frac{3}{4}\cdot\frac{4}{4}=\frac{3}{4}\cdot 1=\frac{3}{4}$$

My Turn!

Perform the indicated operation:

$$\frac{5}{8}+\frac{1}{8}$$

Example 4 Perform the indicated operation:

$$\frac{1}{9}+\frac{2}{15}$$

Since we need a common denominator in order to add or subtract fractions, our first step will be to rewrite each fraction over the lowest common denominator. The lowest common denominator is 45. Then, you follow the same procedure as in Example 1 and add the two numerators together and keep the common denominator.

$$\begin{aligned}\frac{1}{9}+\frac{2}{15}&=\frac{1}{9}\cdot\frac{5}{5}+\frac{2}{15}\cdot\frac{3}{3}\\&=\frac{5}{45}+\frac{6}{45}\\&=\frac{(5+6)}{45}\\&=\frac{11}{45}\end{aligned}$$

My Turn!

Perform the indicated operation:

$$\frac{3}{10}+\frac{5}{12}$$

Subtracting Fractions

Example 5 Perform the indicated operation:

$$\frac{5}{6}-\frac{1}{3}$$

You also need a common denominator when subtracting fractions. So, our first step will be to rewrite each fraction over the lowest common denominator. The lowest common denominator is 6. Once you have rewritten each fraction over the lowest common denominator, subtract the two numerators and keep the common denominator. Be sure that your final answer is reduced!

$$\begin{aligned}\frac{5}{6}-\frac{1}{3}&=\frac{5}{6}-\frac{1}{3}\cdot\frac{2}{2}\\&=\frac{5}{6}-\frac{2}{6}\\&=\frac{(5-2)}{6}\\&=\frac{3}{6}\\&=\frac{1}{2}\end{aligned}$$

My Turn!

Perform the indicated operation:

$$\frac{11}{15}-\frac{5}{12}$$

Since there are several topics in statistics that require subtracting fractions from 1, we will cover that type of subtraction problem here.

Example 6 Perform the indicated operation:

$$1-\frac{3}{8}$$

The number 1 can always be rewritten as $\frac{a}{a}, a \neq 0$. We will select a as the value of the common denominator. So, we set $a=8$ for this problem. Then

$$1-\frac{3}{8}=\frac{8}{8}-\frac{3}{8}=\frac{5}{8}$$

My Turn!

Perform the indicated operation:

$$1-\frac{16}{19}$$

Multiplying Fractions

Example 7 Perform the indicated operation:

$$\frac{5}{6}\cdot\frac{1}{3}$$

You do *not* need a common denominator in order to multiply fractions.

One way to multiply fractions is to multiply all numerators together and make that product the numerator, multiply all denominators together and make that product the denominator, and then reduce the fraction. Another way to do it is to "cancel out" any common factors that are in any numerator with any denominator (recall that you can do this because any number divided by itself is equal to 1) and then multiply across numerators and denominators. However, you do not have to decide on your preferred method for this particular example as there are no common factors in the numerators and denominators.

$$\frac{5}{6}\cdot\frac{1}{3}=\frac{5\cdot 1}{6\cdot 3}=\frac{5}{18}$$

My Turn!

Perform the indicated operation:

$$\frac{4}{15}\cdot\frac{5}{12}$$

Dividing Fractions

Example 8 Perform the indicated operation:

$$\frac{5}{6} \div \frac{1}{3}$$

You do *not* need a common denominator in order to divide fractions.

To divide, multiply by the reciprocal of the second fraction.

(Perhaps, you have heard the expression, "Keep. Change. Flip." This pneumonic device helps some individuals remember to keep the first fraction the same, to change the division to multiplication, and to flip the divisor.)

Then, you follow the steps for multiplication. That is, multiply across the numerators, multiply across the denominators, and simplify.

$$\frac{5}{6} \div \frac{1}{3} = \frac{5}{6} \cdot \frac{3}{1} = \frac{5 \cdot 3}{6 \cdot 1} = \frac{15}{6} = \frac{5}{2}$$

You do not have to rewrite as a mixed number. Mixed numbers are generally not used in statistics.

Also, notice that you could have canceled out the common factor of 3 from the 3 in the second numerator and from the 6 in the first denominator before multiplying.

$$\frac{5}{6} \cdot \frac{3}{1} = \frac{5 \cdot 1}{2 \cdot 1} = \frac{5}{2}$$

My Turn!

Perform the indicated operation:

$$\frac{4}{9} \div \frac{7}{12}$$

Example 9 Perform the indicated operation:

$$\left(\frac{5}{6}\right)^3$$

Recall that an exponent is simply shorthand for repeated multiplication.

$$\left(\frac{5}{6}\right)^3 = \frac{5}{6}\cdot\frac{5}{6}\cdot\frac{5}{6} = \frac{125}{216}$$

My Turn!

Perform the indicated operation:

$$\left(\frac{2}{5}\right)^4$$

Objective 4: Apply operations to decimals.

We will review the basic operations of addition, subtraction, multiplication, division, and exponentiation of decimals. Although you may typically compute these using a calculator within a statistics course, we will briefly review how to do them without technology. Then, you may want to try these examples again using your calculator.

Adding Decimals

Example 10 Perform the indicated operation:

$$1.26+0.8$$

In order to add decimals, you need to line up the decimal points vertically. Then, add as you would normally and vertically line up the decimal point in the sum with those in the addends.

$$\begin{array}{r} 1.26 \\ \underline{+0.80} \\ 2.06 \end{array}$$

My Turn!

Perform the indicated operation: $6.005+1.34$

Subtracting Decimals

Example 11 Perform the indicated operation:

$$1.26 - 0.8$$

You also need to line up the decimals points vertically when you are subtracting decimals.

$$\begin{array}{r} 1.26 \\ \underline{-0.80} \\ 0.46 \end{array}$$

Borrowing was involved in the subtraction. Since you will be using technology in this course, I encourage you to familiarize yourself with the procedure for performing basic operations on your preferred technology.

My Turn!

Perform the indicated operation:

$$9 - 1.004$$

Multiplying Decimals

Example 12 Perform the indicated operation:

$$3.628 \cdot 0.2$$

You do *not* line up decimal points for multiplication.

Count the number of places to the right of each decimal point in each factor. After completing standard multiplication, you will have to insert the decimal point in the answer. The product will have the same number of digits to the right of the decimal point as the sum of the number of places to the right of the decimal point from the factors.

3.628 has three digits to the right of the decimal point.

0.2 has one digit to the right of the decimal point.

The product will have $3+1=4$ digits to the right of the decimal point.

$$\begin{array}{r} 3.628 \\ \times\ 0.2 \\ \hline 0.7256 \end{array}$$

My Turn!

Perform the indicated operation:

$$5.2 \cdot 31.57$$

Dividing Decimals

Example 13 Perform the indicated operation:

$$3.628 \div 0.2$$

First, we will set up the values for long division.

$$0.2\overline{)3.628}$$

Then we must move the decimal point in the divisor so that it is at the right-hand side of the divisor. In this case, we would need to move it one place value to the right (equivalent to multiplying by 10). We also need to do this same movement to the decimal point in the dividend. (You can think about why we are allowed to move the decimal point the same number of places in the divisor and dividend.)

So, we now have

$$2\overline{)36.28}$$

Perform long division.

$$\begin{array}{r} 18.14 \\ 2\overline{)36.28} \\ \underline{-2} \\ 16 \\ \underline{-16} \\ 02 \\ \underline{-2} \\ 08 \\ \underline{-8} \\ 0 \end{array}$$

Make sure you line up the decimal point in your quotient directly above the one in the dividend.

Even though we have briefly reviewed the procedure here, you are strongly encouraged to calculate these operations with technology.

My Turn!

Perform the indicated operation:

$$34.6953 \div 5.1$$

Objective 5: Determine the intersection, union, and complement of two sets.

A set is simply a collection of items. We will review three operations on sets.

The **intersection** of two sets consists of all the elements that are simultaneously in both sets. The symbol that we use to indicate intersection is $\cap$.

Let A be a set and let B be a set; then $A \cap B$ consists of the elements that are in A **and** in B.

The **union** of two sets consists of all the elements that are in at least one of the sets. The symbol that we use to indicate union is $\cup$.

Let A be a set and let B be a set; then $A \cup B$ consists of the elements that are in A **or** in B.

The **complement** of a set A consists of the items that are in the universal set, U (the set that contains everything or all elements under consideration), that are **not** in A. Several symbols are commonly used to denote the complement; however, we will use $\overline{A}$ (read A-bar).

Example 14 Let $U = \{1,2,3,4,5,6,7,8,9\}, A = \{2,4,6,8\}$, and $B = \{1,2,3,4,5\}$ .U = { 1, 2, 3, 4, 5, 6, 7, 8, 9}, A = {2, 4, 6, 8}, and B ={1, 2, 3, 4, 5}.

a) Find $A \cap B$.
To find the intersection, we must look for the elements in A and in B.
$A = \{\boxed{2},\boxed{4},6,8\}$, and $B = \{1,\boxed{2},3,\boxed{4},5\}$ Those elements are 2 and 4.

$$A \cap B = \{2,4\}$$

b) Find $A \cup B$.
To find the union, we must look for the elements that are in A or in B (That is, join them all together). We only need to write each item once, though, in our solution set.

$$A \cup B = \{1,2,3,4,5,6,8\}$$

c) Find $\overline{A}$.
To find the complement of A, we must identify the elements in the universal set that are not in A. The elements in A are $A = \{2,4,6,8\}$. The elements that are in the universal set that are not in A are boxed: $U = \{\boxed{1},2,\boxed{3},4,\boxed{5},6,\boxed{7},8,\boxed{9}\}$. So, the complement of A consists of the elements 1, 3, 5, 7 and 9.

$$\overline{A} = \{1,3,5,7,9\}$$

My Turn!

Let $U = \{1,2,3,4,5,6,7,8,9\}$, $A = \{1,2,3,4,5,6\}$, and $B = \{3,6\}$.

Find $A \cap B$, $A \cup B$, and $\overline{A}$.

Chapter 6 Modeling Random Events: The Normal and Binomial Models

Objective 1: Find area.

In the chapter in the text on normal distributions, we will be finding area and associating it with probability. For the topic of uniform distributions, you will be finding area under rectangles. So, we will take a moment to review finding the area of a rectangle.

Example 1 Find the area of the rectangle shown.

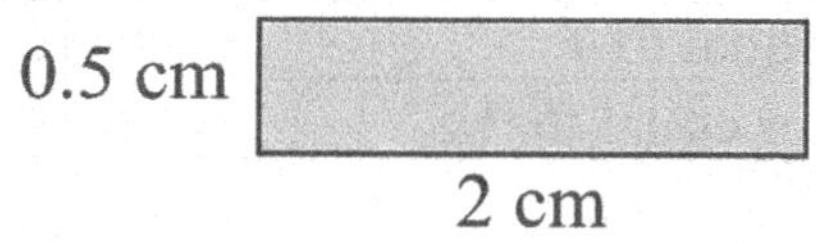

The area of a rectangle is given by $\text{base} \times \text{height}$ (or $\text{length} \times \text{width}$). We can write this as a formula.

$$A = b \cdot h$$

For the above rectangle, the base is 2 cm and the height is 0.5 cm.

We can plug these values into the formula.

$$A = b \cdot h = 2\,\text{cm} \cdot 0.5\,\text{cm} = 1\,\text{cm}^2$$

My Turn!

Find the area of the shaded portion of the rectangle.

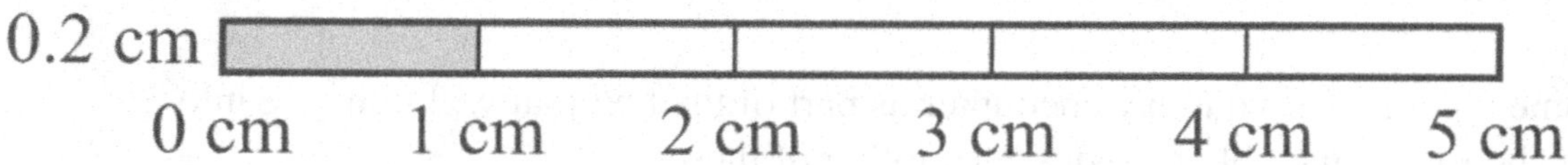

Objective 2: Interpret inequality notation.

In the related section of the textbook, we will be dealing with translating intervals in context to notation that we can use for calculating purposes. Also, you will want to be comfortable with indicating which integers would be located within an interval. The following table includes a column for inequality notation, a column with the inequality expressed in words, and a third column that indicates which integers from 0 to 10 satisfy the inequality from the first column.

Notation	**Translation**	**$X = \{0, 1, 2, 3, 4, 5, 6, 7, 8, 9, 10\}$**
$x < 5$	x is less than 5	0, 1, 2, 3, 4
$x > 7$	x is greater than 7	8, 9, 10
$2 < x < 6$	x is greater than 2 and less than 6	3, 4, 5
$x \leq 4$	x is less than or equal to 4	0, 1, 2, 3, 4
$x \geq 9$	x is greater than or equal to 9	9, 10

Example 2 Which integers from 0 to 10 satisfy the following inequality?

$$x < 7$$

This translates to "x is less than 7." The integers that are less than 7 include 0, 1, 2, 3, 4, 5, and 6.

My Turn!

Which integers from 5 to 15 satisfy the following inequality?

x is greater than or equal to 10.

There are some phrases that you may encounter as part of the text that will translate into inequalities. The following table includes a sampling of these.

Phrase	**Inequality**
At least 5 puppies	$x \geq 5$
No more than 5 puppies	$x \leq 5$
Fewer than 5 puppies	$x < 5$
More than 5 puppies	$x > 5$
Exactly 5 puppies	$x = 5$

There are other possibilities beyond the phrases used in the above table. The more problems you practice, the more comfortable you will be become with the various ways to describe inequalities.

Example 3 Translate the following phrases into inequalities.

a. A heart rate of at least 80 bpm

We can locate the phrase of *at least* in the table and see that it implies ≥. If a heart rate is at least 80 pbm, it means that it can be 80, 81, 82,...bpm.

Letting h represent heart rate, we can translate the phrase into $h \geq 80$.

b. A speed of no more than 55 mph

The phrase *no more than* in the second row of the table and see that it translates to ≤. If the speed is no more than 55mph, it means that it can be anywhere from 0 mph up to and including 55 mph.
If we let s represent speed, the phrase can be rewritten as $s \leq 55$.

c. A temperature below 98.6 degrees Fahrenheit

The phrase *below* is not contained within the table. However, if we think about it, a temperature below 98.6 degrees means that it can be any value less than 98.6, not including 98.6.
Therefore, using t for temperature, our translation becomes $t < 98.6$.

My Turn!

Translate the following phrases into inequalities.

a. A weight of no more than 180 pounds.
b. A height of at least 60 inches
c. Exactly 9 doors in a house

Objective 3: Evaluating formulas for normal probability distributions.

You will be learning about several formulas related to the normal distribution in the corresponding section of the textbook. Here, we will practice evaluating these formulas. You will learn the meaning of the formulas and how to apply them in context within the text section itself.

Example 4 Evaluate $z = \dfrac{x-\mu}{\sigma}$ if $x = 21.3$, $\mu = 23.5$, and $\sigma = 2.1$. Round the answer to the nearest hundredth.

First, we must substitute for the variables in the correct location. Then, we must simplify the numerator. (This is because you can envision the entire numerator within parentheses, and the order of operations dictates that parentheses must be simplified first.) Next, we will divide, and then we will round our final answer.

$$z = \frac{21.3-23.5}{2.1} = \frac{-2.2}{2.1} \approx -1.047 \approx -1.05$$

My Turn!

Evaluate $z = \dfrac{x-\mu}{\sigma}$ if $x = 10.3$, $\mu = 10.2$, and $\sigma = 1.3$. Round the answer to the nearest hundredth.

Example 5 Evaluate $x = \mu + z \cdot \sigma$ if $\mu = 6.7$, $z = 1.96$, and $\sigma = 0.9$. Round the answer to the nearest tenth.

First, we must substitute for the variables in the correct location. Then, we will follow the order of operations for the right-hand side. So, we will have to multiply before adding.

$$x = \mu + z \cdot \sigma = 6.7 + 1.96 \cdot 0.9 = 6.7 + 1.764 = 8.464 \approx 8.5$$

My Turn!

Evaluate $x = \mu + z \cdot \sigma$ if $\mu = 9.8$, $z = 2.575$, and $\sigma = 0.3$. Round the answer to the nearest tenth.

Example 6 Solve for x when $z = \frac{x - \mu}{\sigma}$ and $z = 2.15$, $\mu = 3.5$, and $\sigma = 0.3$. Round the answer to the nearest tenth.

We need to substitute 2.15 for z to the left of the equal sign, and 3.5 for μ and 0.3 for σ on the right-hand side.

$$2.15 = \frac{x - 3.5}{0.3}$$

Now, we need to solve for x. The first step is to multiply both sides of the equation by 0.3 to eliminate the fraction.

$$2.15(0.3) = \frac{(x - 3.5)}{0.3}(0.3)$$

$$0.645 = x - 3.5$$

Add 3.5 to both sides to isolate x.

$$0.645 + 3.5 = x - 3.5 + 3.5$$

$$4.145 = x$$

Now, we must round our final answer to the nearest tenth and $x \approx 4.1$.

My Turn!

Solve for x when $z = \frac{x-\mu}{\sigma}$ and $z = -0.34$, $\mu = 21.5$, and $\sigma = 0.8$. Round the answer to the nearest tenth.

Example 7 Evaluate

$$z = \frac{\overline{x}-\mu}{\frac{\sigma}{\sqrt{n}}}$$

if $\overline{x} = 9.8$, $\mu = 25.6$, $\sigma = 1.9$, and $n = 49$. Round your answer to the nearest hundredth.

We need to substitute 9.8 for $\overline{x}$ and 25.6 for μ in the numerator, and 1.9 for σ and 49 for n in the denominator.

$$z = \frac{9.8-25.6}{\frac{1.9}{\sqrt{49}}}$$

We can then simplify the right-hand side using the order of operations. You may want to visualize an invisible set of parentheses around the numerator and another invisible set of parentheses around the denominator.

$$\begin{aligned} z &= \frac{9.8-25.6}{\frac{1.9}{\sqrt{49}}} \\ &= \frac{(9.8-25.6)}{\left(\frac{1.9}{\sqrt{49}}\right)} \\ &= \frac{-15.8}{\left(\frac{1.9}{\sqrt{49}}\right)} \\ &= \frac{-15.8}{\left(\frac{1.9}{7}\right)} \\ &= \frac{-15.8}{0.27142857} \\ &= -58.2105266 \end{aligned}$$

Finally, we must round our final answer to the hundredth, which gives us $z \approx -58.21$.

My Turn!

Evaluate

$$z = \frac{\overline{x}-\mu}{\frac{\sigma}{\sqrt{n}}}$$

if $\overline{x} = 7.2, \mu = 5.8, \sigma = 1.2,$ and $n = 81$. Round your answer to the nearest hundredth.

Objective 4: Evaluate formulas for binomial distributions.

Example 8 Evaluate $\mu + 2\sigma$ when $\mu = 25.2$ and $\sigma = 2.3$.

We need to substitute 25.2 for μ (the lowercase Greek letter mu) in the beginning of the expression and 1.3 for σ (the lowercase Greek letter sigma) at the end of the expression.

$\mu + 2\sigma = 25.2 + 2(2.3)$

Following the order of operations, we need to multiply before we add.

$25.2 + 2(2.3) = 25.2 + 4.6 = 29.8$

My Turn!

Evaluate $\mu + 2\sigma$ when $\mu = 9.9$ and $\sigma = 0.4$.

Example 9 Evaluate $\mu = n \cdot p$ given $n = 150$ and $p = 0.35$.

We are evaluating a formula given the values for all but one variable, μ.

$$\mu = n \cdot p = 150 \cdot 0.35 = 52.5$$

My Turn!

Evaluate $\mu = n \cdot p$ when $n = 2000$ and $p = 0.04$.

Example 10 Evaluate $\sigma = \sqrt{np(1-p)}$ $n = 50$ and $n = 0.2$. Round the answer to the nearest tenth.

Begin by replacing *n* and *p* with the given values. You will be substituting 0.2 for *p* in two places. After you have done this, follow the order of operations.

$$\sigma = \sqrt{np(1-p)} = \sqrt{50 \cdot 0.2 \cdot (1-0.2)} = \sqrt{50 \cdot 0.2 \cdot (0.8)} = \sqrt{10 \cdot 0.8} = \sqrt{8} \approx 2.82 \approx 2.8$$

My Turn!

Evaluate $\sigma = \sqrt{np(1-p)}$ for $n = 100$ and $p = 0.38$. Round the answer to the nearest tenth.

Example 11 Evaluate $45 \cdot p^x \cdot (1-p)^{n-x}$ for $n = 10$, $x = 2$, and $p = 0.3$. Round to the nearest thousandth.

First, we must substitute the given values in for the given variables.

$$45 \cdot p^x \cdot (1-p)^{n-x} = 45 \cdot 0.3^2 \cdot (1-0.3)^{10-2}$$

Next, we will clean up the third factor.

$$= 45 \cdot 0.3^2 \cdot (0.7)^8$$

Per the order of operations, we will evaluate exponents next.

$$= 45 \cdot 0.09 \cdot 0.05764801$$

Next, we multiply.

$$= 0.2334744405$$

Finally, we round to the nearest thousandth.

$$\approx 0.233$$

My Turn!

Evaluate $15{,}504 \cdot p^x \cdot (1-p)^{n-x}$ for $n = 20$, $x = 5$, and $p = 0.4$. Round to the nearest ten thousandth.

Chapter 7 Survey Sampling and Inference

Objective 1: Find the middle value for an interval given either in interval notation or as an inequality.

One topic that you will have to understand in the text is how to find the middle value for a given interval. The given interval can be written either as an inequality or in interval notation. We will review how to handle both of these formats.

Example 1 Let $3.8 < \mu < 6.2$ represent an interval on the number line. Find the value that is in the middle of the interval.

From an algebra class, you may recall that this is the same thing as finding the midpoint for two points on a number line. In order to find the middle value, you have to add the two values of the endpoints and divide that sum by 2.

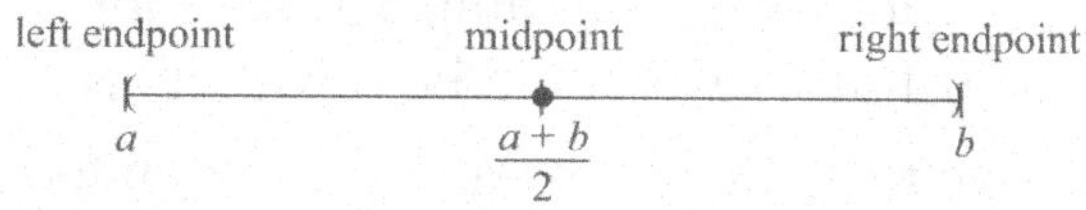

$$\text{The middle value of } a < \mu < b \text{ is } \frac{a+b}{2}.$$

We have $3.8 < \mu < 6.2$. So, substitute 3.8 for a and 6.2 for b. Then, the middle value of the given interval of numbers is found by following the order of operations. (Recall that the entire numerator can be thought of as being within parentheses and is therefore simplified first.)

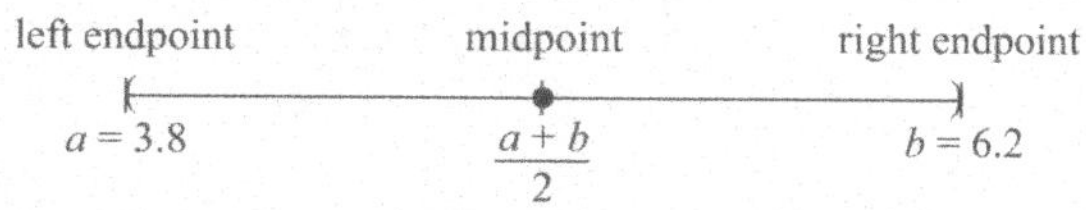

$$\frac{a+b}{2} = \frac{3.8+6.2}{2} = \frac{10}{2} = 5$$

My Turn!

Let $105.67 < \mu < 118.59$ represent an interval on the number line. Find the value that is in the middle of the interval.

Example 2 If $(0.124, 0.498)$ is an interval on the number line, find the value that is in the middle of the interval.

This problem is completed in the same way as Example 1. It simply differs in how the given interval is written. The interval $(0.124, 0.498)$ could be rewritten as $0.124 < \mu < 0.498$. In order to find the middle value, you have to add the two values of the endpoints and divide that sum by 2. That is,

$$\text{the middle value of } (a,b) = \frac{a+b}{2}$$

We have $(0.124, 0.498) = (a, b)$. So, substitute 0.124 for a and 0.498 for b and simplify.

$$\frac{a+b}{2} = \frac{0.124 + 0.498}{2} = \frac{0.622}{2} = 0.311$$

My Turn!

If $(6.48, 15.84)$ is an interval on the number line, find the value that is in the middle of the interval.

Objective 2: Find the distance from the middle value of an interval to its endpoints.

Example 3 Let $(0.368, 0.549)$ represent an interval on the number line. Find the distance from the middle of the interval to either endpoint.

Let m represent the distance from the midpoint to an endpoint.

There are two ways to solve this problem:

Method 1: Find the middle value of the interval first and then find the distance from the midpoint to the endpoint.

Step 1: Follow the process for Objective 1 to find the middle value of the interval.

The middle value of $(0.368, 0.549)$ is

$$\frac{0.368+0.549}{2}=\frac{0.917}{2}=0.4585$$

Step 2: Find the distance from the middle value that you found in step 1 to the endpoint.

You can find the distance from the middle value to an endpoint by subtracting the left endpoint from the midpoint of the interval.

distance from middle value to endpoint = middle value − left endpoint

(You can also calculate this distance by subtracting the middle value from the right endpoint.)

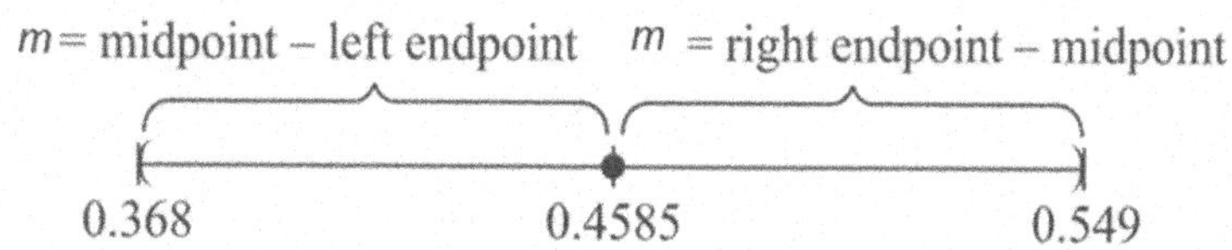

For this example, the distance from the middle value to either endpoint is

$$m = 0.4585 - 0.368 = 0.0905$$

Method 2: Find the distance directly without actually finding the midpoint.

The distance from the middle value to the endpoint of an interval (a, b) can be given by

$$m = \frac{b-a}{2}$$

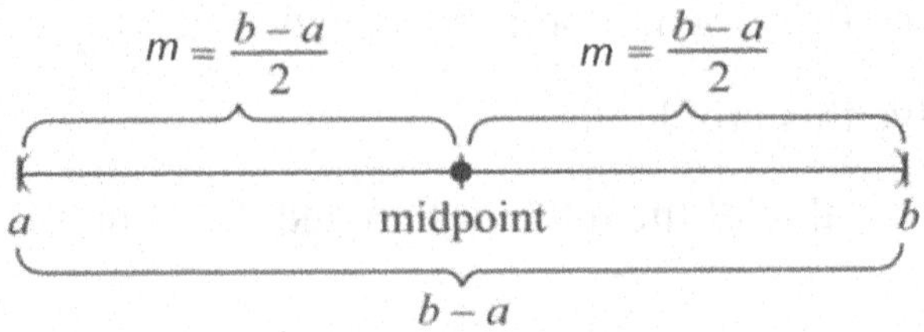

So, the distance from the middle value to the endpoint of the interval $(0.368, 0.549)$ can be found as follows:

$$m = \frac{0.549 - 0.368}{2} = \frac{0.181}{2} = 0.0905$$

You can feel free to use the method that is more intuitive for you. Sometimes you may need to know the middle value anyway. In that case, method 1 is definitely the way to go. If you don't need to know the midpoint, then you may want to use method 2.

My Turn!

Let $(26.8, 32.4)$ represent an interval on the number line. Find the distance from the middle of the interval to either endpoint.

Objective 3: Write and interpret three different forms of intervals (as used for confidence intervals).

As you may have gathered from the examples above, there are several different forms of intervals. We will review the various forms you can see in statistics. Often, technology will give you the output in a form that is different from how you may want to write it. So, it is important to understand how to translate from one format to another.

The key point that we have to remember is that $a < p < b$ and (a, b) represent the same interval for p. Note that p is a variable expression that you will use quite a bit in statistics.

Example 4: Express $2.5 < p < 5.3$ in interval notation.

The endpoints of the interval are 2.5 and 5.3. So, we can rewrite $2.5 < p < 5.3$ as $(2.5, 5.3)$.

My Turn!

Express $-3.5 < p < 1.9$ in interval notation.

Example 5 Express $(0.88, 0.96)$ in inequality notation, assuming that the variable under study is p.

The endpoints of the interval are 0.88 and 0.96. So, we can rewrite $(0.88, 0.96)$ as $0.88 < p < 0.96$.

My Turn!

Express $(0.63, 0.92)$ in inequality notation, assuming that the variable under study is p.

Sometimes, you will want to write your interval in the following form:

$$\text{middle value}(\hat{p}) \pm \text{distance from middle value to endpoint}(m) = \hat{p} \pm m$$

Note that $\hat{p}$ is read as *p-hat.* The notation $\pm$ translates to *plus or minus*.

Example 6 Let $4.0 < p < 4.8$ represent an interval on the number line. Write the given interval in the format $\hat{p} \pm m$.

First, we must find the value that is in the middle of the interval and let the variable $\hat{p}$ represent that.

$$\hat{p} = (4.0, 4.8) = \frac{4.0+4.8}{2} = \frac{8.8}{2} = 4.4$$

Next, we must find the distance from the middle of the interval to either endpoint and let m represent that.

$$m = 4.4 - 4.0 = 0.4$$

Finally, we must write the given interval in the desired format $\hat{p} \pm m$.

So, $4.0 < p < 4.8$ can be rewritten in the desired form of 4.4 ± 0.4.

My Turn!

Let $0.18 < p < 0.36$ represent an interval on the number line. Write the given interval in the format $\hat{p} \pm m$.

Objective 4: Evaluate formulas used for confidence intervals for proportions.

You will be learning about several formulas in the related section of the textbook on confidence intervals. Here, we will practice evaluating these formulas. You will learn the meaning of the formulas and how to apply them in context within the text section itself.

Example 7 Evaluate the formula

$$m = z^{*} \cdot \sqrt{\frac{\hat{p}(1-\hat{p})}{n}}$$

when $z^{*} = 1.96$, $\hat{p} = 0.516$, and $n = 1000$. Round the answer to the thousandths.

First, we must substitute for the variables in the correct location. Then, we will follow the order of operations for the right-hand side.

$$\begin{aligned} m &= z^{*} \cdot \sqrt{\frac{\hat{p}(1-\hat{p})}{n}} \\ &= 1.96 \cdot \sqrt{\frac{0.516(1-0.516)}{1000}} \\ &= 1.96 \cdot \sqrt{\frac{0.516(0.484)}{1000}} \\ &= 1.96 \cdot \sqrt{\frac{0.249744}{1000}} \\ &= 1.96 \cdot \sqrt{0.000249744} \\ &= 1.96 \cdot 0.0158032908 \\ &= 0.03097445 \\ &\approx 0.0310 \end{aligned}$$

My Turn!

Evaluate the formula $m = z^* \cdot \sqrt{\frac{\hat{p}(1-\hat{p})}{n}}$ when $z^* = 1.645$, $\hat{p} = 0.874$, and $n = 500$. Round the answer to the thousandths.

Objective 5: Evaluate formulas used for confidence intervals for the difference of two proportions.

Example 8 Evaluate

$$m = z^* \cdot \sqrt{\frac{\hat{p}_1 \cdot (1-\hat{p}_1)}{n_1} + \frac{\hat{p}_2 \cdot (1-\hat{p}_2)}{n_2}}$$

if $z^* = 1.96$, $x_1 = 50$, $x_2 = 60, n_1 = 70, n_2 = 100,$ $\hat{p}_1 = \frac{x_1}{n_1}$, and $\hat{p}_2 = \frac{x_2}{n_2}$.

Round your answer to the nearest thousandth.

Let's begin by using the formulas for the variables that we need to plug into the formula for *m*. Let's find numerical values for $\hat{p}_1$, and $\hat{p}_2$. Throughout this solution, we will round to the hundred-thousandth (two more places than that desired our final answer is normally adequate) in intermediate steps to minimize rounding error.

$$\hat{p}_1 = \frac{x_1}{n_1} = \frac{50}{70} \approx 0.71429$$

$$\hat{p}_2 = \frac{x_2}{n_2} = \frac{60}{100} = 0.6$$

$$1 - \hat{p}_1 \approx 1 - 0.71429 = 0.28571$$

$$1 - \hat{p}_2 = 1 - 0.6 = 0.4$$

Since we now have numbers for each of the variables on the right-hand side, we can plug them into the formula for *m*. Then, we follow the order of operations on the right-hand side to simplify it.

$$\begin{aligned}
m &= z^* \cdot \sqrt{\frac{\hat{p}_1 \cdot (1-\hat{p}_1)}{n_1} + \frac{\hat{p}_2 \cdot (1-\hat{p}_2)}{n_2}} \\
&\approx 1.96\sqrt{\frac{0.71429 \cdot 0.28571}{70} + \frac{0.6 \cdot 0.4}{100}} \\
&\approx 1.96\sqrt{\frac{0.20408}{70} + \frac{0.24}{100}} \\
&\approx 1.96\sqrt{0.0029154 + 0.0024} \\
&= 1.96\sqrt{0.0053154} \\
&\approx 1.96 \cdot 0.072907 \\
&\approx 0.14290 \\
&\approx 0.143
\end{aligned}$$

My Turn!

Evaluate

$$m = z^* \cdot \sqrt{\frac{\hat{p}_1 \cdot (1-\hat{p}_1)}{n_1} + \frac{\hat{p}_2 \cdot (1-\hat{p}_2)}{n_2}}$$

if $z^* = 2.575$, $x_1 = 32$, $x_2 = 33$, $n_1 = 50$, $n_2 = 60$, $\hat{p}_1 = \frac{x_1}{n_1}$, and $\hat{p}_2 = \frac{x_2}{n_2}$.

Round your answer to the nearest thousandth.

Chapter 8 Hypothesis Testing for Population Proportions

Objective 1: Evaluate formulas for hypothesis testing for proportions.

Example 1 Evaluate the formula

$$z = \frac{\hat{p} - p_0}{\sqrt{\frac{p_0(1-p_0)}{n}}}$$

when $\hat{p} = \frac{x}{n}$, $x = 80$, $n = 100$, and $p_0 = 0.75$. Round your answer to the nearest hundredth.

First we need to figure out the value for $\hat{p}$ so that we can substitute in that value. Since, $\hat{p} = \frac{x}{n}$, with $x = 80$ and $n = 100$, we get

$$\hat{p} = \frac{x}{n} = \frac{80}{100} = 0.8$$

Now, we can substitute in values for all of the variables and solve for z.

$$z = \frac{\hat{p} - p_0}{\sqrt{\frac{p_0(1-p_0)}{n}}} = \frac{0.8 - 0.75}{\sqrt{\frac{0.75 \cdot (1-0.75)}{100}}} = \frac{0.05}{\sqrt{\frac{0.75 \cdot 0.25}{100}}} = \frac{0.05}{\sqrt{\frac{0.1875}{100}}} = \frac{0.05}{\sqrt{0.001875}} \approx \frac{0.05}{0.0433} \approx 1.1547$$

Now, we will round our value for z to the nearest hundredth.

$$z \approx 1.15$$

My Turn!

Evaluate the formula

$$z = \frac{\hat{p} - p_0}{\sqrt{\frac{p_0(1-p_0)}{n}}}$$

when $\hat{p} = \frac{x}{n}$, $x = 90$, $n = 120$, and $p_0 = 0.35$. Round your answer to the nearest hundredth.

Objective 2: Evaluate formulas for two-sample hypothesis testing for proportions.

Example 2 Evaluate the formula

$$\hat{p} = \frac{x_1 + x_2}{n_1 + n_2}$$

for $x_1 = 15$, $x_2 = 18$, $n_1 = 35$, and $n_2 = 25$. Round your answer to the nearest ten thousandth.

First, we substitute the given values in for the variables in the formula. Notice that the variables with *x* and *n* have subscripts. The variables x_1 and x_2 represent two different unknowns. So, it is critical that we substitute the appropriate values into the variables with the correct subscripts.

$$\hat{p} = \frac{x_1 + x_2}{n_1 + n_2} = \frac{15 + 18}{35 + 25}$$

Once we have completed substituting, we apply the order of operations to the right-hand side. We can group the numerator and the denominator in two sets of parentheses to remind us to individually simplify each of them first by adding. Then, we divide the numerator by the denominator to change the fraction to a decimal. No rounding of the final answer is necessary in this problem.

$$\hat{p} = \frac{(15 + 18)}{(35 + 25)} = \frac{33}{60} = 0.55$$

My Turn!

Evaluate the formula

$$\hat{p} = \frac{x_1 + x_2}{n_1 + n_2}$$

for $x_1 = 22$, $x_2 = 18$, $n_1 = 40$, and $n_2 = 30$. Round your answer to the nearest ten thousandth.

Example 3 Evaluate the formula

$$z = \frac{(\hat{p}_1 - \hat{p}_2) - 0}{\sqrt{\hat{p}(1-\hat{p})\left(\frac{1}{n_1} + \frac{1}{n_2}\right)}}$$

if $x_1 = 40$, $x_2 = 60, n_1 = 80, n_2 = 100,\ \hat{p}_1 = \frac{x_1}{n_1},\ \hat{p}_2 = \frac{x_2}{n_2},\ \hat{p} = \frac{x_1 + x_2}{n_1 + n_2}$.

Round your answer to the nearest hundredth.

We have a lot of variables to substitute for in the formulas. We even have formulas to find out the values for some of these variables! Let's begin by using the formulas for the variables that we need to plug into the formula for z. We will find numerical values for $\hat{p}_1$, $\hat{p}_2$, and $\hat{p}$.

$$\hat{p}_1 = \frac{x_1}{n_1} = \frac{40}{80} = 0.5$$

$$\hat{p}_2 = \frac{x_2}{n_2} = \frac{60}{100} = 0.6$$

$$\hat{p} = \frac{x_1 + x_2}{n_1 + n_2} = \frac{40+60}{80+100} = \frac{100}{180} \approx 0.5556$$

As you can see, we round $\hat{p}$ to the nearest ten thousandth even though we are rounding our final answer to the nearest hundredth. We keep a few extra places in intermediate steps to minimize rounding error. Ideally, we would try to complete all the calculations at one time using technology so that there is no intermediate rounding. However, when that is not possible, an extra two places should generally be sufficient.

Now, we are ready to substitute into the formula for z and simplify. Note that entire expression under the radical symbol can be thought of as being inside parentheses. This means we cannot take a square root until we have simplified the entire expression under the $\sqrt{\ }$.

$$\begin{aligned}
z &= \frac{(\hat{p}_1 - \hat{p}_2) - 0}{\sqrt{\hat{p}(1-\hat{p})\left(\frac{1}{n_1} + \frac{1}{n_2}\right)}} \\
&\approx \frac{(0.5 - 0.6) - 0}{\sqrt{0.5556 \cdot 0.4444\left(\frac{1}{80} + \frac{1}{100}\right)}} \\
&= \frac{-0.1}{\sqrt{0.5556 \cdot 0.4444\left(\frac{1}{80} + \frac{1}{100}\right)}} \\
&\approx \frac{-0.1}{\sqrt{0.2469\left(\frac{1}{80} + \frac{1}{100}\right)}} \\
&\approx \frac{-0.1}{\sqrt{0.2469(0.0125 + 0.01)}} \\
&= \frac{-0.1}{\sqrt{0.2469(0.0225)}} \\
&\approx \frac{-0.1}{\sqrt{0.0056}} \\
&\approx \frac{-0.1}{0.0748} \\
&\approx -1.3369 \\
&\approx -1.34
\end{aligned}$$

My Turn!

Evaluate the formula

$$z = \frac{(\hat{p}_1 - \hat{p}_2) - 0}{\sqrt{\hat{p}(1-\hat{p})\left(\frac{1}{n_1} + \frac{1}{n_2}\right)}}$$

if $x_1 = 40$, $x_2 = 30, n_1 = 60, n_2 = 80$, $\hat{p}_1 = \frac{x_1}{n_1}$, $\hat{p}_2 = \frac{x_2}{n_2}$, and $\hat{p} = \frac{x_1 + x_2}{n_1 + n_2}$.

Round your answer to the nearest hundredth.

Chapter 9 Inferring Population Means

Objective 1: Evaluate formulas used for confidence intervals for means.

You will be learning about several formulas in the related section of the textbook on confidence intervals. Here, we will practice evaluating these formulas. You will learn the meaning of the formulas and how to apply them in context within the text section itself.

Example 1 Evaluate the formula

$$m = t^* \cdot \frac{s}{\sqrt{n}}$$

when $t^* = 2.626$, $s = 0.93$, and $n = 200$. Round your answer to the nearest hundredth.

By now, you probably have the idea that we need to substitute and simplify!

$$m = t^* \cdot \frac{s}{\sqrt{n}} = 2.626 \cdot \frac{0.93}{\sqrt{200}} \approx 2.626 \cdot \frac{0.93}{14.14213562} \approx 0.1726882039 \approx 0.17$$

My Turn!

Evaluate the formula

$$m = t^* \cdot \frac{s}{\sqrt{n}}$$

when $t^* = 2.756$, $s = 0.64$, and $n = 30$. Round your answer to the nearest hundredth.

Objective 2: Evaluate formulas for hypothesis testing for means.

Example 2 Evaluate the formula

$$t = \frac{\overline{x} - \mu}{\frac{s}{\sqrt{n}}}$$

when $\mu = 99.2$, $\overline{x} = 99.5$, $n = 100$, and $s = 1.3$. Round your answer to the nearest hundredth.

Begin by substituting the given values for the variables in the formula. Then, follow the order of operations to simplify the right-hand side.

$$t = \frac{\overline{x} - \mu}{\frac{s}{\sqrt{n}}} = \frac{99.5 - 99.2}{\frac{1.3}{\sqrt{100}}} = \frac{0.3}{\frac{1.3}{\sqrt{100}}} = \frac{0.3}{\frac{1.3}{10}} = \frac{0.3}{0.13} \approx 2.3077 \approx 2.31$$

Note that if 100 were not a perfect square, you would want to use your calculator to find the value of the square root.

My Turn!

Evaluate the formula

$$t = \frac{\overline{x} - \mu}{\frac{s}{\sqrt{n}}}$$

when $\mu = 16$, $\overline{x} = 17.5$, $n = 80$, and $s = 2.4$. Round your answer to the nearest hundredth.

Objective 3: Evaluate formulas used for confidence intervals for comparing two means.

Example 3 Evaluate $m = t^* \cdot \sqrt{\frac{s_1^{\ 2}}{n_1} + \frac{s_2^{\ 2}}{n_2}}$ if $t^* = 2.449, s_1 = 1.5, s_2 = 1.8, n_1 = 50, \text{and } n_2 = 40$. Round your answer to the nearest tenth.

By now, you know the routine; substitute and simplify!

$$\begin{aligned} m &= t^* \cdot \sqrt{\frac{s_1^{\ 2}}{n_1} + \frac{s_2^{\ 2}}{n_2}} \\ &= 2.449 \cdot \sqrt{\frac{1.5^2}{50} + \frac{1.8^2}{40}} \\ &= 2.449 \cdot \sqrt{\frac{2.25}{50} + \frac{3.24}{40}} \\ &= 2.449 \cdot \sqrt{0.045 + 0.081} \\ &= 2.449 \cdot \sqrt{0.126} \\ &\approx 2.449 \cdot 0.3550 \\ &\approx 0.8694 \\ &\approx 0.9 \end{aligned}$$

My Turn!

Evaluate $m = t^* \cdot \sqrt{\frac{s_1^{\ 2}}{n_1} + \frac{s_2^{\ 2}}{n_2}}$ if $t^* = 1.99, s_1 = 0.6, s_2 = 0.8, n_1 = 30, \text{and } n_2 = 50$. Round your answer to the nearest tenth.

Objective 4: Evaluate formulas for hypothesis testing for comparing two population means.

Example 4 Evaluate the formula

$$t = \frac{(\bar{x}_1 - \bar{x}_2) - (\mu_1 - \mu_2)}{\sqrt{\frac{s_1^{\,2}}{n_1} + \frac{s_2^{\,2}}{n_2}}}$$

for $\bar{x}_1 = 210, \bar{x}_2 = 190,\ \mu_1 - \mu_2 = 0,\ s_1 = 11,\ s_2 = 7,\ n_1 = 45$, and $n_2 = 60$. Round your answer to the nearest hundredth.

We can begin this problem by immediately substituting the given values for the variables. Once again, we use the order of operations to guide us in simplifying the right-hand side. Be careful not to evaluate the square root until you have simplified the entire expression under the $\sqrt{\ }$.

$$\begin{aligned}
t &= \frac{(\bar{x}_1 - \bar{x}_2) - (\mu_1 - \mu_2)}{\sqrt{\frac{s_1^{\,2}}{n_1} + \frac{s_2^{\,2}}{n_2}}} \\
&= \frac{(210-190)-(0)}{\sqrt{\frac{11^2}{45} + \frac{7^2}{60}}} \\
&= \frac{20}{\sqrt{\frac{11^2}{45} + \frac{7^2}{60}}} \\
&= \frac{20}{\sqrt{\frac{121}{45} + \frac{49}{60}}} \\
&\approx \frac{20}{\sqrt{2.6889 + 0.8167}} \\
&= \frac{20}{\sqrt{3.5056}} \\
&\approx \frac{20}{1.8723} \\
&\approx 10.6820 \\
&\approx 10.68
\end{aligned}$$

My Turn!

Evaluate the formula

$$t = \frac{(\bar{x}_1 - \bar{x}_2) - (\mu_1 - \mu_2)}{\sqrt{\frac{s_1^2}{n_1} + \frac{s_2^2}{n_2}}}$$

for $\bar{x}_1 = 68, \bar{x}_2 = 64$, $\mu_1 - \mu_2 = 0$, $s_1 = 4$, $s_2 = 3$, $n_1 = 80$, and $n_2 = 90$.

Round your answer to the nearest hundredth.

Example 5 Evaluate

$$t = \frac{\overline{x}_{difference} - \mu_{difference}}{\frac{s_{difference}}{\sqrt{n}}}$$

for $\overline{x}_{difference} = 1.2, \mu_{difference} = 0,\ s_{difference} = 0.14$, and $n = 49$. Round your answer to the nearest hundredth.

Let's substitute the four given values in for the appropriate variables. Be careful that you follow the order of operations carefully, especially for the denominator of the complex fraction. If you are using your calculator, you may have to insert parentheses.

$$t = \frac{\overline{x}_{difference} - \mu_{difference}}{\frac{s_{difference}}{\sqrt{n}}} = \frac{1.2-0}{\frac{0.14}{\sqrt{49}}} = \frac{1.2}{\frac{0.14}{\sqrt{49}}} = \frac{1.2}{\frac{0.14}{7}} = \frac{1.2}{0.02} = 60$$

My Turn!

Evaluate

$$t = \frac{\overline{x}_{difference} - \mu_{difference}}{\frac{s_{difference}}{\sqrt{n}}}$$

for $\overline{x}_{difference} = 0.9, \mu_{difference} = 0,\ s_{difference} = 0.4$, and $n = 64$. Round your answer to the nearest hundredth.

Chapter 10 Associations between Categorical Variables

Objective 1: Evaluate formulas for testing with categorical variables.

Example 1 Evaluate the formula $E = \frac{r \bullet c}{g}$ for $r = 30, c = 50,$ and $g = 98$. Round the answer to the nearest hundredth.

After all the other formulas we have looked at, this one seems simple by comparison! Simply substitute the three values and simplify,

$$E = \frac{r \bullet c}{g} = \frac{30 \bullet 50}{98} = \frac{1500}{98} \approx 15.306 \approx 15.31$$

My Turn!

Evaluate the formula $E = \frac{r \bullet c}{g}$ for $r = 45, c = 37,$ and $g = 100$. Round the answer to the nearest hundredth.

Example 2 Find $\chi^2 = \Sigma\frac{(O-E)^2}{E}$ for the given table of values.

O	E
34	35
48	45
25	28
76	73

Round the answer to the nearest hundredth.

Note that χ^2 (chi-squared) is considered as an entity. You do not approach this problem trying to solve for χ. You strive to get a value for χ^2. That is, $\chi^2 = \square$.

This notation is telling us to subtract each E from each corresponding O value first since we work out parentheses first in the order of operations. Next we would calculate exponents per the order of operations. Then, we sum up these squares of these differences. Once we have this sum, then we can perform division by the denominator.

$$\begin{aligned}
\chi^2 &= \Sigma\frac{(O-E)^2}{E} \\
&= \frac{(34-35)^2}{35} + \frac{(48-45)^2}{45} + \frac{(25-28)^2}{28} + \frac{(76-73)^2}{73} \\
&= \frac{(-1)^2}{35} + \frac{(3)^2}{45} + \frac{(-3)^2}{28} + \frac{(3)^2}{73} \\
&= \frac{1}{35} + \frac{9}{45} + \frac{9}{28} + \frac{9}{73} \\
&\approx 0.0286 + 0.2 + 0.3214 + 0.1233 \\
&= 0.6733 \\
&\approx 0.67
\end{aligned}$$

My Turn!

Find $\chi^2 = \Sigma \frac{(O-E)^2}{E}$ for the accompanying table. Round the answer to the nearest hundredth.

O	E
25	27
43	45
25	29
18	21